MATLAB Guide to Fibonacci Numbers and the Golden Ratio

A Simplified Approach, for Students and Scientists and Engineers

Peter I. Kattan

Petra Books
www.PetraBooks.com

Peter I. Kattan, PhD

Correspondence about this book may be sent to the author at one of the following two email addresses:

info@petrabooks.com

pkattan@petrabooks.com

MATLAB Guide to Fibonacci Numbers and the Golden Ratio: A Simplified Approach, for Students and Scientists and Engineers.
Written by Peter I. Kattan.
ISBN: 979-8-8691-0251-5

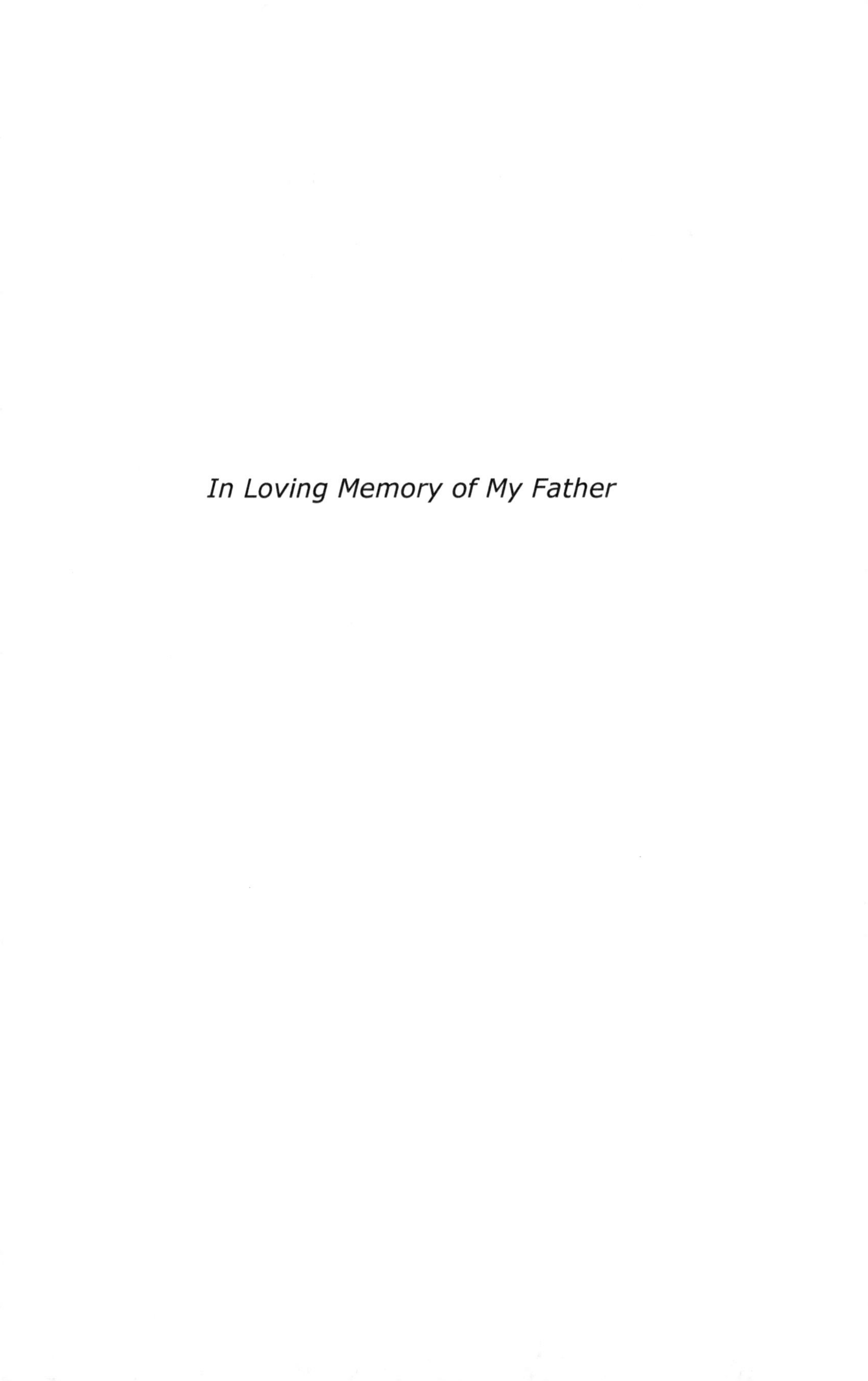

In Loving Memory of My Father

MATLAB Guide to Fibonacci Numbers and the Golden Ratio

A Simplified Approach, for Students and Scientists and Engineers

Preface

This book is for people who love MATLAB and Fibonacci numbers. The book is divided into two major parts. The first part comprises of five chapters that review the essentials of MATLAB that are needed. The second part consists of six chapters on Fibonacci numbers and the Golden Ratio. The book is really designed for beginners and students. The material presented has been simplified to such a degree such that the book may also be used for students and teachers in high schools. One of the objectives of writing this book is to introduce both MATLAB and Fibonacci numbers to high school students.

Chapter 1 provides an overview of MATLAB and may be skipped upon a first reading of the book. Chapter 2 reviews the important topic of symbolic computing with MATLAB. In this chapter, it is shown how to make algebraic calculations using MATLAB. Solving various types of algebraic equations is illustrated in Chapter 3. Next, writing MATLAB functions is presented in Chapter 4. Chapter 5 presents different methods of plotting both two-dimensional and three-dimensional graphs in MATLAB. Thus the first five chapters are a short review of the capabilities of MATLAB that are needed to study Fibonacci numbers.

The second part of the book starts with Chapter 6 on Fibonacci numbers. This is followed by deriving the Golden Ratio in Chapter 7 in great detail and simplification. Various properties of the Golden Ratio are illustrated in Chapter 8. This is followed in Chapter 9 by studying other types of sequences that are related to the Fibonacci sequence. In particular, the Lucas sequence is presented in this chapter. Chapter 10 presents various generalizations of Fibonacci numbers while Chapter 11 concludes the presentation with a short overview of random Fibonacci numbers.

The material presented in this book has been tested with version 7 of MATLAB and should work with any prior or later versions. The material presented is very easy and simple to understand – written in a simplified manner. An extensive references list is also provided at the end of the book with references to books and numerous web links for more information. The references provided will guide you to other resources where you can get more information.

I would like to thank my family members for their help and continued support without which this book would not have been possible. In this edition, I am providing two email addresses for my readers to contact me – pkattan@petrabooks.com
info@petrabooks.com

January 2024

Peter I. Kattan

Contents

MATLAB Guide to Fibonacci Numbers and the Golden Ratio
A Simplified Approach, for Students and Scientists and Engineers

Part I

Review of MATLAB

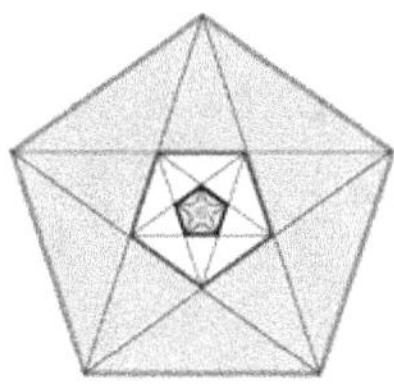

MATLAB Guide to Fibonacci Numbers and the Golden Ratio

A Simplified Approach, for Students and Scientists and Engineers

1. Introduction

In this introductory chapter a short MATLAB tutorial is provided. This tutorial describes the basic MATLAB commands needed. Note that there are numerous free MATLAB tutorials on the internet – check references [1-28]. Also, you may want to consult many of the excellent books on the subject – check references [29-47]. This chapter may be skipped on a first reading of the book.

In this tutorial it is assumed that you have started MATLAB on your computer system successfully and that you are now ready to type the commands at the MATLAB prompt (which is denoted by double arrows ">>"). For installing MATLAB on your computer system, check the web links provided at the end of the book.

Entering scalars and simple operations is easy as is shown in the examples below:

```
>> 2*3+5

ans =

    11
```

The order of the operations will be discussed in subsequent chapters.

```
>> cos(60*pi/180)

ans =

    0.5000
```

The argument for the `cos` command and the value of `pi` will be discussed in subsequent chapters.

We assign the value of 4 to the variable x as follows:

```
>> x = 4

x =

        4

>> 3/sqrt(2+x)

ans =

    1.2247
```

To suppress the output in MATLAB use a semicolon to end the command line as in the following examples. If the semicolon is not used then the output will be shown by MATLAB:

```
>> y = 30;
>> z = 8;
>> x = 2; y-z;
>> w = 4*y + 3*z

w =

    144
```

MATLAB is case-sensitive, i.e. variables with lowercase letters are different than variables with uppercase letters. Consider the following examples using the variables x and X:

```
>> x = 1

x =

        1
>> X = 2
```

```
X =

    2

>> x

x =

    1

>> X

X =

    2
```

Use the `help` command to obtain help on any particular MATLAB command. The following example demonstrates the use of `help` to obtain help on the `det` command which calculated the determinant of a matrix:

```
>> help det
 DET    Determinant.
    DET(X) is the determinant of the square
matrix X.

    Use COND instead of DET to test for
matrix singularity.

    See also cond.

    Overloaded functions or methods (ones
with the same name in other directories)
        help sym/det.m
    Reference page in Help browser
        doc det
```

The following examples show how to enter matrices and perform some simple matrix operations:

```
>> x = [1 2 3 ; 4 5 6 ; 7 8 9]

x =

        1        2        3
        4        5        6
        7        8        9

>> y = [1 ; 0 ; -4]

y =

        1
        0
       -4

>> w = x*y

w =

      -11
      -20
      -29
```

Let us now see how to use MATLAB to solve a system of simultaneous algebraic equations. Let us solve the following system of simultaneous algebraic equations:

$$\begin{bmatrix} 3 & 5 & -1 \\ 0 & 4 & 2 \\ -2 & 1 & 5 \end{bmatrix} \begin{Bmatrix} x_1 \\ x_2 \\ x_3 \end{Bmatrix} = \begin{Bmatrix} 2 \\ 1 \\ -4 \end{Bmatrix} \tag{1.1}$$

We will use Gaussian elimination to solve the above system of equations. This is performed in MATLAB by using the backslash operator "\" as follows:

```
>> A= [3 5 -1 ; 0 4 2 ; -2 1 5]

A =

     3      5     -1
     0      4      2
    -2      1      5

>> b = [2 ; 1 ; -4]

b =

     2
     1
    -4

>> x = A\b

x =

   -1.7692
    1.1154
   -1.7308
```

It is clear that the solution is $x_1 = -1.7692$, $x_2 = 1.1154$, and $x_3 = -1.7308$. Alternatively, one can use the inverse matrix of A to obtain the same solution directly as follows:

```
>> x = inv(A)*b

x =

   -1.7692
    1.1154
```

```
   -1.7308
```

It should be noted that using the inverse method usually takes longer than using Gaussian elimination especially for large systems of equations.

Consider now the following 4 x 4 matrix D:

```
>> D = [1 2 3 4 ; 2 4 6 8 ; 3 6 9 12 ; -5 -3 -1 0]

D =

    1      2      3      4
    2      4      6      8
    3      6      9     12
   -5     -3     -1      0
```

We can extract the sub-matrix in rows 2 to 4 and columns 1 to 3 as follows:

```
>> E = D(2:4, 1:3)

E =

    2      4      6
    3      6      9
   -5     -3     -1
```

We can extract the third column of D as follows:

```
>> F = D(1:4, 3)

F =

    3
    6
    9
```

```
        -1
```

We can extract the second row of D as follows:

```
>> G = D(2, 1:4)

G =

    2       4       6       8
```

We can also extract the element in row 3 and column 2 as follows:

```
>> H = D(3,2)

H =

    6
```

We will now show how to produce a two-dimensional plot using MATLAB. In order to plot a graph of the function $y = f(x)$, we use the MATLAB command $plot(x,y)$ after we have adequately defined both vectors x and y. The following is a simple example to plot the function $y = x^2 + 3$ for a certain range:

```
>> x = [1 2 3 4 5 6 7 8 9 10 11]

x =
  Columns 1 through 10

      1       2       3       4       5       6       7
8       9      10

  Column 11

     11
```

```
>> y = x.^2 + 3

y =

  Columns 1 through 10

      4       7      12      19      28      39      52
67      84     103

  Column 11

   124

>> plot(x,y)
```

Figure 1.1 shows the plot obtained by MATLAB. It is usually shown in a separate window. In this figure no titles are given to the x- and y-axes. These titles may be easily added to the figure using the `xlabel` and `ylabel` commands. More details about these commands will be presented in the chapter on graphs.

In the calculation of the values of the vector y, notice the dot "." before the exponentiation symbol "^". This dot is used to denote that the operation following it be performed element by element. More details about this issue will be discussed in subsequent chapters.

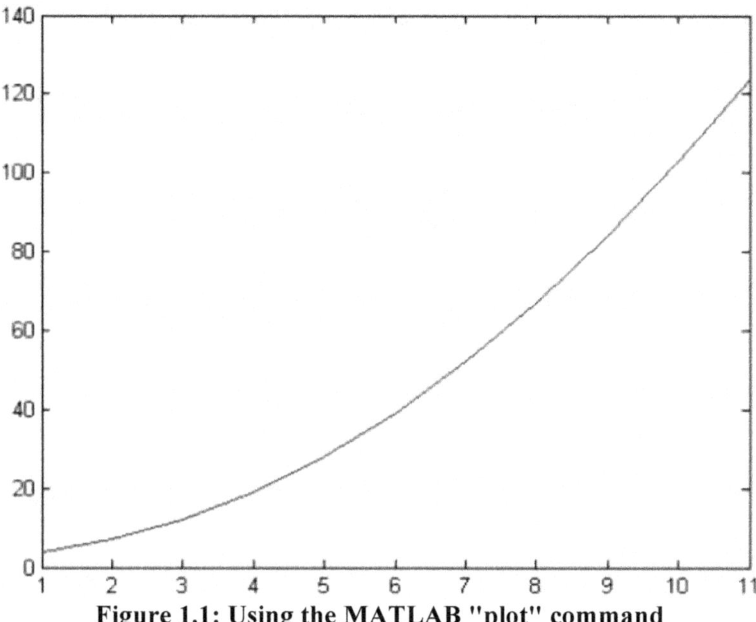

Figure 1.1: Using the MATLAB "plot" command

Finally, we will show how to make magic squares with MATLAB. A magic square is a square grid of numbers where the total of any row, column, or diagonal is the same. Magic squares are produced by MATLAB using the command `magic`. Here is a simple example to produce a 5 x 5 magic square:

```
>> magic(5)

ans =

    17    24     1     8    15
    23     5     7    14    16
     4     6    13    20    22
    10    12    19    21     3
    11    18    25     2     9
```

It is clear that the total of each row, column or diagonal in the above matrix is 65.

MATLAB Guide to Fibonacci Numbers and the Golden Ratio

A Simplified Approach, for Students and Scientists and Engineers

2. Symbolic Computing

In this chapter, the basics of symbolic computing are illustrated with MATLAB. Symbolic computing means performing algebraic and other mathematical operations without resorting to numerical computations. This type of computing is also called computer algebra. Computer programs and software that are used in symbolic computations are called computer algebra systems. Many such competing systems are available today such as MAPLE, MATHEMATICA, MACZYMA, etc. It should be noted that MATLAB is not basically a computer algebra system but can be used for this purpose with the aid of the MATLAB Symbolic Math Toolbox. This toolbox uses the MAPLE engine and helps in converting MATLAB into a computer algebra system.

Some of the operations that we will use in subsequent chapters can be easily implemented using symbolic computing. Thus the use of the MATLAB Symbolic Math Toolbox will become essential in subsequent chapters. Therefore, the use of this toolbox with various symbolic computing exercises will be illustrated briefly in this chapter.

Symbolic arithmetic operations may also be performed in MATLAB using the MATLAB Symbolic Math Toolbox. For the next examples, this toolbox needs to be installed with MATLAB. In order to perform symbolic operations in MATLAB, we need to define symbolic numbers[1] using the sym command. For example, ½ is defined as a symbolic number as follows:

```
>> sym(1/2)

ans =
```

---

[1] Actually these are symbolic objects in MATLAB but objects are beyond the scope of this book.

1/2

In the above example, the number ½ is stored in MATLAB without any approximation using decimal digits like 0.5. Once used with the `sym` command, the computations will be performed algebraically or symbolically. For example, the following addition of two fractions is performed numerically as follows without the `sym` command:

```
>> 1/2+3/5

ans =

    1.1000
```

Adding parentheses will not affect the numerical computation in this case:

```
>> (1/2)+(3/5)

ans =

    1.1000
```

However, using the `sym` command of the MATLAB Symbolic Math Toolbox will result in the computation performed symbolically without the use of decimal digits;

```
>> sym((1/2)+(3/5))

ans =

11/10
```

Notice in the above example how the final answer was cast in the form of a fraction without decimal digits and without calculating a numerical value. In the addition of the two fractions above,

MATLAB finds their common denominator and adds them by the usual procedure for rational numbers. Notice also in the above four example outputs that symbolic answers are not indented but numerical answers are indented.

You may also use symbolic variables[2] in MATLAB. For example, the variables x and y are defined as symbolic variables as follows:

```
>> x=1/2

x =

    0.5000

>> y= 2/3

y =

    0.6667

>> sym(x)

ans =

1/2


>> sym(y)

ans =

2/3
```

---

[2] Actually these are symbolic objects in MATLAB but objects are beyond the scope of this book.

Arithmetic operations can now be performed on these symbolic variables. Consider the following example where the variables x and y were defined as above :

```
>>  z=sym(x-y)

z =

-1/6
```

In the above example, the variable z is used automatically as a symbolic variable. For symbolic operations, the important thing is to start each operation or calculation with the sym command. Here is another example to calculate the volume of a sphere:

```
>>  r=15/7

r =

    2.1429

>>  sym(r)

ans =

15/7


>>  volume = sym(4*pi*r^3/3)

volume =

4500*pi/343
```

In the above example, the final answer is given in the form of a fraction, $\dfrac{4500\pi}{343}$, with no decimal expressions and without calculating a numerical value (no units are used in this example). In case you need to have this value calculated numerically, then you can need to use the $\texttt{double}$ command as follows:

```
>> double(volume)

ans =

    41.2162
```

It is seen thus that the value of the volume of the sphere is 41.2162 (no units are used in this example). Finally, the variables $\texttt{x}$ and $\texttt{this}$ can be defined as symbolic variables without assigning a numerical value as follows:

```
>> x = sym('x')

x =

x


>> this = sym('this')

this =

this
```

Next, we study how to solve algebraic equations using MATLAB.

MATLAB Guide to Fibonacci Numbers and the Golden Ratio
A Simplified Approach, for Students and Scientists and Engineers

3. Solving Equations

In this chapter we discuss how to solve algebraic equations using MATLAB. We will discuss both linear and nonlinear algebraic equations. We will also discuss systems of simultaneous linear and nonlinear algebraic equations. First, let us solve the following simple linear equation for the variable x:

$$2x - 3 = 0$$

The solution of the above equation may be obvious to the reader but we will show how to solve it using MATLAB. The easiest method to do it in MATLAB would be to consider the left-hand-side of the equation as a polynomial of first degree and then find the roots of the polynomial using the MATLAB command roots. First, we write the above equation as a polynomial as follows:

$$p(x) = 2x - 3 = 0$$

We enter the coefficients of the above polynomial in a vector in MATLAB then use the roots command to find the root which is the solution to the above linear equation in x. Here are the needed commands:

```
>> p = [2 -3]

p =

     2    -3

>> roots(p)

ans =

    1.5000
```

It is clear that the correct solution for x is obtained which is 1.5. Next, let us show similarly how to find the solution of a quadratic equation in x by solving the following quadratic equation:

$$5x^2 + 3x - 4 = 0$$

We will consider the above equation as a polynomial of second degree in x. We will enter the coefficients of the polynomial in a vector in MATLAB followed by the `roots` command. Here are the needed MATLAB commands:

```
>> p = [5 3 -4]

p =

      5      3      -4

>> roots(p)

ans =

    -1.2434
     0.6434
```

It is clear from the above that two real solutions are obtained for x. Let us now solve a more complicated equation. Let us solve the following equation for x:

$$2x^5 - 3x^4 - 5x^3 + x^2 - 1 = 0$$

First, we consider the left-hand-side as a polynomial of fifth degree in x. Then we write the coefficients of this polynomial as a vector in MATLAB followed by the `roots` command as follows (note that we will enter 0 for the missing x in the above polynomial):

```
>> p = [2 -3 -5 1 0 -1]
```

```
p =

     2    -3    -5     1     0    -1

>> roots(p)

ans =

   2.4507
  -1.0000
  -0.6391
   0.3442 + 0.4480i
   0.3442 - 0.4480i
```

Note from the above that five values are obtained as the solution of the above equation. These values are the five roots of the polynomial given above. Note that three of these roots are real while two roots are complex conjugates of each other.

Next, let us discuss how to solve a system of linear simultaneous algebraic equations. In order to solve such systems there are several methods available with most of them involving the use of matrices and vectors. Consider the following simple system of two linear algebraic equations in the variables x and y:

$$x - 3y = 5$$
$$4x + 6y = 3$$

In order to solve the above system of equations, we need to re-write the system as a matrix equation. The coefficients of x and y on the left-hand-side of the equations are considered as the elements of a square matrix (of size 2x2 here) while the numbers on the right-hand-side are entered into a vector. The resulting equivalent matrix equation is written as follows:

$$\begin{bmatrix} 1 & -3 \\ 4 & 6 \end{bmatrix} \begin{Bmatrix} x \\ y \end{Bmatrix} = \begin{Bmatrix} 5 \\ 3 \end{Bmatrix}$$

The above matrix equation is in the form $[A]\{x\} = \{b\}$. The solution[3] of this equation is $\{x\} = [A]^{-1}\{b\}$. Thus we need to find the inverse of the coefficient matrix $[A]$ and multiply it with the constant vector $\{b\}$. Here are the needed MATLAB commands along with the solution:

```
>> A = [1 -3 ; 4 6]

A =

      1      -3
      4       6

>> b = [5 ; 3]

b =

      5
      3

>> x = inv(A)*b

x =

    2.1667
   -0.9444
```

Thus it is clear from the above that x = 2.1667 and y = -0.9444 are the solutions to the above system of equations. Note that in order to solve the above system we had to find the inverse of

---

[3] Consult a book on linear algebra for the proof of this solution.

a 2x2 matrix. In this case, it was quick to find this inverse because the coefficient matrix was small. But for larger matrices, finding the inverse using MATLAB may take more time. Therefore, it is advised to use another method to solve the above system. One such method that is rather fast in execution is called Gaussian elimination. This method is already implemented in MATLAB as matrix division using the backslash operator "\". Here is the solution of the above system again using Gaussian elimination and the backslash operator:

```
>> A = [1 -3 ; 4 6]

A =

        1       -3
        4        6

>> b = [5 ; 3]

b =

        5
        3

>> x = A\b

x =

    2.1667
   -0.9444
```

It is clear that we obtain exactly the same solution as before but with more speed on the part of MATLAB. The use of Gaussian elimination along with the backslash operator in MATLAB is greatly recommended for solving large systems of algebraic linear simultaneous equations. As an example, let us solve the following system of five algebraic linear simultaneous equations:

$$2x_1 - 4x_2 - x_3 + 3x_4 - x_5 = 3$$
$$x_1 + x_2 - 2x_3 + x_5 = 6$$
$$-x_1 - 3x_2 + x_4 + 3x_5 = -4$$
$$3x_1 - x_2 - x_3 + 4x_4 - x_5 = 1$$
$$x_1 + x_2 - x_3 + 2x_4 = 5$$

First, we re-write the above system as a matrix equation as follows:

$$\begin{bmatrix} 2 & -4 & -1 & 3 & -1 \\ 1 & 1 & -2 & 0 & 1 \\ -1 & -3 & 0 & 1 & 3 \\ 3 & -1 & -1 & 4 & -1 \\ 1 & 1 & -1 & 2 & 0 \end{bmatrix} \begin{bmatrix} x_1 \\ x_2 \\ x_3 \\ x_4 \\ x_5 \end{bmatrix} = \begin{Bmatrix} 3 \\ 6 \\ -4 \\ 1 \\ 5 \end{Bmatrix}$$

The following are the MATLAB commands using Gaussian elimination along with the solution:

```
>> A = [2 -4 -1 3 -1 ; 1 1 -2 0 1 ; -1 -3 0
1 3 ; 3 -1 -1 4 -1 ; 1 1 -1 2 0]

A =

    2    -4    -1     3    -1
    1     1    -2     0     1
   -1    -3     0     1     3
    3    -1    -1     4    -1
    1     1    -1     2     0

>> b = [3 ; 6 ; -4 ; 1 ; 5]

b =

    3
```

```
     6
    -4
     1
     5

>> x = A\b

x =

   -4.3571
    0.3571
   -6.4286
    1.2857
   -2.8571
```

It is seen from the above result that five real solutions are obtained for the above system of linear equations.

Systems of simultaneous nonlinear algebraic equations can also be solved in MATLAB but these are significantly more difficult to solve. The reason is that there are no direct commands to solve these complicated systems in MATLAB. One will have to use the MATLAB Optimization Toolbox to find certain functions for the solution of these types of equations — for example the command `fsolve` in this toolbox will solve a system of nonlinear equations. Another option is to use the MATLAB Symbolic Math Toolbox to solve such nonlinear systems - using the MATLAB command `solve`. This option will be illustrated in detail below.

Solving Equations with the MATLAB Symbolic Math Toolbox

The MATLAB Symbolic Math Toolbox can be used to solve algebraic equations in MATLAB. In addition, systems of linear and nonlinear algebraic equations can also be solved using this toolbox. There are special commands in this toolbox for solving equations — in particular the MATLAB command `solve`. Please note that the command `roots` that was discussed earlier in this chapter cannot be used for the symbolic solution of algebraic equations. It needs to be replaced with the `solve` command. The use of the MATLAB command `solve` will be illustrated in this section with several examples.

Consider the following linear algebraic symbolic equation that we need to solve for the variable x in terms of the constants a and b:

$$ax + b = 0$$

We will use the MATLAB command `solve` to solve the above equation. Note that we cannot use the trick of writing it as a polynomial and use the `roots`[4] command. Here is the needed format to use the `solve` command in order to solve the above equation:

```
>> syms x
>> syms a
>> syms b
>> x = solve('a*x+b = 0')

x =

-b/a
```

It is noted that the correct solution is obtained which is $x = -\dfrac{b}{a}$. Next, let us solve the following quadratic equation for x in terms of the constants a, b, and c:

$$ax^2 + bx + c = 0$$

---

[4] The `roots` command can only be used with numerical computations.

The following is the correct format of the `solve` command in order to obtain the two solutions of the above nonlinear equation:

```
>> syms x
>> syms a
>> syms b
>> syms c
>> x = solve('a*x^2 + b*x + c = 0')


x =


 1/2/a*(-b+(b^2-4*a*c)^(1/2))
 1/2/a*(-b-(b^2-4*a*c)^(1/2))
```

It is clear that we obtain the correct two solutions[5] of the quadratic formula. Let us now solve another nonlinear equation not involving a polynomial. Consider the following equation:

$$e^x = \sin x$$

The above equation is a highly nonlinear equation in x. Below is the needed format of the `solve` command to find this solution:

```
>> syms x
>> x = solve('exp(x)-sin(x) = 0')


x =


.36270205612105111082838863639609-
1.13374591941375253711670812190577*i
```

It is clear from the above that we obtained one complex solution of the above equation. The resulting solution is

---

[5] We can use the MATLAB command `pretty` to display the solution in a nice way but this will not be done here. The correct use of the `pretty` command in the example above would be `pretty(x)`.

approximated as $0.363 - 1.13i$. At this step, one may use the MATLAB command `format short` to display the solution using four decimal digits only.

Next, we will consider systems of equations. Consider the following system of linear simultaneous algebraic equations:

$$ax - by = c$$
$$2ax + b = 9c$$

The above system of linear equations can also be solved for the variables x and y in terms of the constants a, b, and c using the MATLAB `solve` command as follows:

```
>> syms x
>> syms y
>> syms a
>> syms b
>> syms c
>> [x,y] = solve('a*x - b*y - c = 0', '2*a*x
+ b - 9*c = 0')

x =

-1/2/a*(b-9*c)




y =

-1/2*(-7*c+b)/b
```

It is clear that two solutions are obtained for x and y in terms of a, b, and c as shown above. Our final example will illustrate the solution of a nonlinear system of simultaneous algebraic equations. Consider the following system:

$$x^3 - xy = -3$$
$$x^2 - 2y^2 = 5$$

The above system is a highly nonlinear system. Its solution using the MATLAB command `solve` is illustrated below:

```
>> syms x
>> syms y
>> [x,y] = solve('x^3 - x*y + 3 = 0', 'x^2 -
2*y^2 - 5 = 0')

x =

1.1146270688593392975541180081115+1.36934983
263003820407393290850 74*i

.36979370624989001773301777958373+1.06182311
95136687011590573146631*i
 -
1.4844207751092293152871357876952+.282946327
302944510326135 34596558*i
 -1.4844207751092293152871357876952-
.282946327302944510326135 34596558*i
  .36979370624989001773301777958373-
1.0618231195136687011590573146631*i
  1.1146270688593392975541180081115-
1.3693498326300382040739329085074*i


y =
.43988651431894061086040911013399+1.73489563
8424583512029045010 1938*i
 -.11319578028438869117106079077457-
1.734408764002893871669123 8072308*i
```

```
    .17330926596544808031065168064 07-
1.21173961516022589100765017481 34*i

.17330926596544808031065168064 07+1.211739615
16022589100765017481 34*i
   -
.11319578028438869117106079077457+1.73440876
4002893871669123807230 8*i
   .43988651431894061086040911013399-
1.73489563842458351202904501019 38*i
```


It is clear from the above that we obtain six complex solutions for the above nonlinear system. Finally, one may use the MATLAB command `format short` to display the six solutions above using four decimal digits only.

4. MATLAB Functions

In this chapter we introduce the basics of programming in MATLAB. The material presented in this chapter is not intended to be comprehensive or exhaustive but merely an introduction[6] to programming in MATLAB. In MATLAB programming, the list of commands and instructions are usually stored in a text file called an M-file (short for MATLAB file). These M-files can be of two kinds – either script files or function files. These files can be created or opened from the `File` menu by clicking on `New` or `Open`, then clicking on `M-File`. These files typically have the `.m` extension that is associated with MATLAB.

We will discuss script files first. The examples presented are simple. The reader can write more complicated examples based on the material presented here. Script files are used to store MATLAB scripts. A script is defined in MATLAB as a sequence of MATLAB commands. A script can span several lines. For example, here is a MATLAB script:

```
% This is an example
cost = 50
profit = 10
sale_price = cost + profit
```

Note that the first line in the script above is a comment line – this is optional. Store the above example in an M-file called `example1.m` then run the example by typing `example1` at the command line to get:

```
>> example1

cost =
```

---

[6] The interested reader may consult the references listed at the end of this book for more details about programming in MATLAB.

```
    50


profit =

    10


sale_price =

    60
```

There are several remarks about the previous example. The first remark is that a script needs to be defined first (in an M-file) then executed by typing the name of the file (without the `.m` extension) on the command line. Note the use of the comment symbol `%` on the first line which is a comment line. The use of comments in scripts is optional but is useful to the reader. Note also that sometimes we do not need the results of all the commands displayed by MATLAB. For this purpose, we can use semicolons to suppress the outputs that are not needed. For example, here is the same example above with the output suppressed with semicolons except the last line:

```
% This is an example
cost = 50;
profit = 10;
sale_price = cost + profit
```

Store the above script in a new M-file and call it `example2.m`. Executing the above example by typing `example2` at the command line will result in the following output:

```
>> example2

sale_price =
    60
```

The second type of M-files in MATLAB is called a function file. These are more structured[7] than script files. In general, scripts are easier to write than functions but functions have their advantages. In fact, many of the commands used by MATLAB are stored in function files. Functions usually contain a sequence of MATLAB commands possibly spanning several lines. In addition, a function has a function name[8] and one or more arguments. In addition, a function returns a value[9]. In contrast, a script does not have any arguments. For example, here is a function called `area(r)` that is stored in an M-File called `area.m`

```
function area(r)
% This is a function
area = pi*r^2
```

The above function calculates the area of a circle of radius r. The first line of the function file must have the word `function` followed by the name of the function and the argument(s) in parentheses. The last line of the function usually includes a calculation of the variable that is used for the name of the function. Now, run the above function by typing the name of the function followed by the argument in parentheses along with its value on the command line. Here is the result:

```
>> area(3)

area =

    28.2743
```

---

[7] This means that there are rules for writing functions in MATLAB.

[8] The name of the function must be exactly the same as the name of the function file it is stored in.

[9] Consult a book on programming for more details about functions and return values. In other programming languages, functions are called procedures or subroutines and have a different structure than in MATLAB.

A function may be executed several times with a different value for the argument each time. For example here is another execution of the previous function with the value 5 replacing 3[10]:

```
>> area(5)

area =

    78.5398
```

A function may be executed as many times as needed. However, it needs to be defined only once in an M-file. Here is an example of a function that calculates the perimeter of a rectangle with sides a and b:

```
function perimeter(a,b)
% This is another example of a function
perimeter = 2*(a+b)
```

It is clear that the above function has two arguments, namely a and b. Now, execute[11] the above function three times, each time for a certain rectangle with a specified length and width as follows:

```
>> perimeter(2,3)

perimeter =

    10

>> perimeter(3,7)

perimeter =
```

---

[10] No units are used in this example.
[11] The two words "execute" and "run" are used interchangeably in this chapter. They are basically equivalent for our purpose in this chapter.

```
    20

>> perimeter(1.2,5.3)

perimeter =

    13
```

It is clear that the first execution above calculates the perimeter for a rectangle with sides 2 and 3, the second execution calculates the perimeter with sides 3 and 7, while the third execution calculates the perimeter with sides 1.2 and 5.3[12].

Next, we will discuss briefly several constructs[13] that are used in programming in MATLAB - inside scripts and functions. In particular, we will discuss loops (the For loop and the While loop) and decisions (the If Elseif construct and the Switch Case construct).

In MATLAB loops, several command lines are executed repeatedly over many cycles depending on certain parameters in the loop. There are two types of loops in programming in MATLAB – the For loop and the While loop. Consider the following script file which is to be stored under the file name example3.m – this is an example of a For loop.

```
% This is an example of a FOR loop
for n = 1:10
    x(n) = n^2;
end
```

Now, run the above script by typing the command example3 at the command prompt as follows:

---

[12] Note that no units are specified for the dimensions used in this example.

[13] It should be noted that these programming constructs are also available in other programming languages but they have a different structure, i.e. different rules for their implementation.

```
>> example3
```

No output will be displayed because of the presence of the semicolon in the script file above. However, to find the value of the variable x, you need only to type x at the command prompt as follows:

```
> x

x =

     1      4      9     16     25     36     49
64     81    100
```

It is seen from the above example that the For loop cycles over the values of n from 1 to 10 – exactly 10 times. In each cycle, the value of the vector x is calculated using the quadratic formula inside the For loop. The final result for the vector x is displayed above – it is seen that the vector x has exactly 10 elements.

Here is another example of a For loop to be stored in a script file called example4.m

```
%This is an example of another FOR loop
for n = 1:3
    for m = 1:3
        y(m,n) = m^2 + m*n + n^2;
    end
end
```

Actually the above example contains two For loops that are nested. Now, run the above script file by typing example4 at the command prompt as follows, then type y to get the values of the elements of the matrix y:

```
>> example4
```

```
>> y

y =

      3      7     13
      7     12     19
     13     19     27
```

It is seen that two nested For loops in this example can produce a matrix with a single assignment inside the For loops.

Another way to use loops in MATLAB is to use the While loop instead of the For loop. Here is an example of a While loop to be stored in a script file called example5.m

```
% This is an example of a While loop
tol = 0.0;
n = 0;
while tol < 10
    n = n + 1;
    tol = tol + 2;
end
```

Now, run the above example by typing example5 at the command prompt, then type tol and n to get their values after executing the script:

```
>> example5
>> n

n =

      5

>> tol

tol =
```

```
10
```

It is seen from the above example that the loop will continue to cycle as long as the value of the variable `tol` is less than[14] 10. As soon as the value of `tol` becomes equal to or larger than 10, the loop ends. The final values of the variables are then displayed by MATLAB with the appropriate commands. Note also the use of the semicolons in the script file.

Next, we discuss implementing decisions in programming in MATLAB. There are several constructs in MATLAB to use for decisions like the `If` construct, the `If Else Then` construct, the `If Elseif Then` construct, and the `Switch Case` construct. We will discuss these constructs briefly in the remaining part of this chapter.

Consider the following script file containing an `If` construct to be stored in a script file called `example6.m`

```
% This is an example using If Then
boxes = 10;
cost = boxes*3;
if boxes > 7
    cost = boxes*2.5;
end
```

Now, run the above example by typing `example6` at the command line. Then type the variable `cost` to see which answer we get as follows:

```
>> example6
>> cost
```

---

[14] The logical operator "less than" is represented in MATLAB by the symbol <. We do not cover logical operators in this book but this is an instance where their use is needed.

```
cost =

    25
```

It is clear from the above example that we get 25 as the value of the variable cost. This is because the statement inside the If construct was executed. Actually, what happens is as follows: First, MATLAB computes the value of the variable cost as 30 using the first assignment statement for the variable cost. However, once we enter the If construct, the value of the variable cost may change depending on certain parameters. In this case, the value will change only if the value of the variable boxes is greater than 7. In our case, the value of the variable boxes is 10, so the effect of the If construct is to change the computation of the value of the variable cost from 30 to 25 using the second assignment statement for the variable cost – the one inside the If construct.

Here is the same example above but written as a function in a function file called cost.m. The function is called cost(boxes) and has one argument only. Writing the example as a function and not as a script will give us the option of changing the value of the variable boxes at the command prompt to see what effect it will have on the value of the variable cost. Here is the definition of the function:

```
function cost(boxes)
% This is an example of an If construct
cost = boxes*3
if boxes > 7
    cost = boxes*2.5
end
```

Now, execute the above function using different values for the variable boxes. In this case, we will execute it three times to see the effect it has on the value of the variable cost as follows:

```
>> cost(4)

cost =

    12

>> cost(6)

cost =

    18

>> cost(8)

cost =

    24


cost =

    20
```

It is seen that the values of the variable cost for the first two executions were obtained directly using the first assignment statement for cost – this is because the value of the variable boxes is less than or equal to 7 in these two cases. However, for the third execution, the value of cost was calculated initially to be 24 but changed to 20 finally and correctly because the value of the variable boxes is larger than 7 in this last case.

Next, we discuss the If Elseif[15] construct using a slightly modified version of the function cost(boxes). In this

---

[15] Note that there other variations of the above constructs used for implementing decisions in MATLAB, e.g. the If Else construct (see the last section in this chapter on the Symbolic Math Toolbox to see an example of using this construct).

case, we call the function `cost2(boxes)` and define it as follows to be stored in a function file called `cost2.m`

```
function cost2(boxes)
% This is an example of an If Elseif construct
if boxes < 7
     cost = boxes*2.5
elseif boxes < 9
     cost = boxes*2
elseif boxes > 9
     cost = boxes*1.5
end
```

Now, we execute the above function several times to test the results we get. We will execute it three times with the values of 6, 8, and 10 for the variable `boxes`. Here is what we get:

```
>> cost2(6)

cost =

    15

>> cost2(8)

cost =
    16

>> cost2(10)

cost =

    15
```

Consult a book dedicated on programming in MATLAB for more details about these constructs.

In the above example, it is seen that the appropriate branch of the `If Elseif` construct and the appropriate assignment statement is used for the computation of the value of the variable `cost` depending on the value of the variable `boxes`. Note that several `Elseif` 's were needed to accomplish this. A more efficient way of doing this is to use the `Switch Case` construct. This is achieved by defining a new function `cost3(boxes)` to be stored in the function file `cost3.m` as follows (note that the functions `cost2` and `cost3` are not exactly equivalent):

```
function cost3(boxes)
% This is an example of the Switch Case construct
switch boxes
    case 6
        cost = boxes* 2.5
    case 8
        cost = boxes* 2
    case 10
        cost = boxes*1.5
    otherwise
        cost = 0
end
```

Now, execute the above function three times as before with the three values of 6, 8, and 10 for the variable `boxes`. This is what we get (exactly as before):

```
>> cost3(6)
cost =

    15

>> cost3(8)

cost =

    16
```

```
>> cost3(10)

cost =

    15
```

In the next chapter, we will discuss graphs and plotting in MATLAB.

MATLAB Guide to Fibonacci Numbers and the Golden Ratio

A Simplified Approach, for Students and Scientists and Engineers

5. Graphs in MATLAB

In this chapter we explain how to plot graphs in MATLAB. Both two-dimensional and three-dimensional graphs are presented. A graph in MATLAB appears in its own window (not on the command line). First, we will consider two-dimensional or planar graphs. To plot a two-dimensional graph, we need two vectors. Here is a simple example using two vectors x and y along with the MATLAB command plot:

```
>> x = [1 2 3 4 5]

x =

      1       2       3       4       5

>> y = [3 9 12 10 6]

y =

      3       9       12      10       6

>> plot(x,y)
```

The resulting graph is displayed in its own window and is shown in Figure 9.1. Note how the command plot was used above along with the two vectors x and y. This is the simplest use of this command in MATLAB.

We can add some information to the above graph using other MATLAB commands that are associated with the plot command. For example, we can use the title command to add a title to the graph. Also, we can use the MATLAB commands xlabel and ylabel to add labels to the x-axis and the y-axis. But before using these three commands (title, xlabel, ylabel), we need to keep the plotted graph in its window. For this we use the `hold on` command as follows:

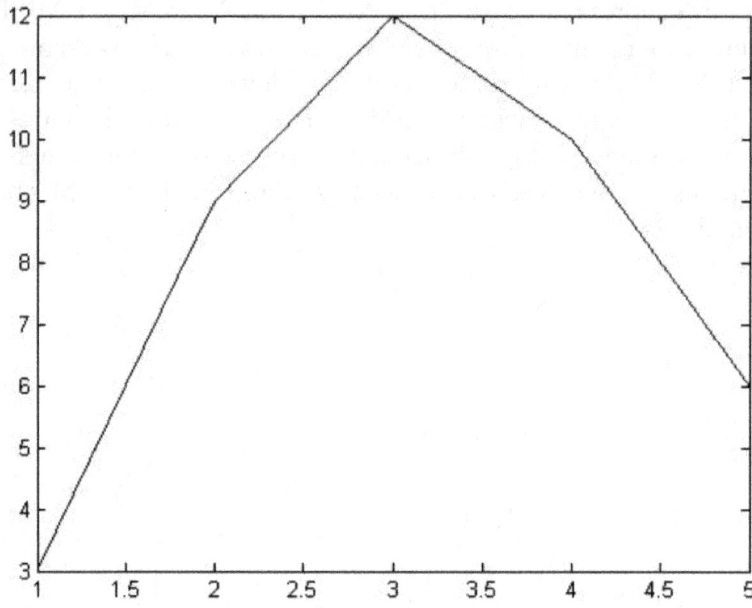

Figure 9.1: A simple use of the plot command

```
>> hold on
>> title('This is an example of a simple graph')
>> xlabel('x')
>> ylabel('y')
```

After executing the above commands, the plotted graph appears as shown in Figure 9.2 with the title and axes labels clearly displayed.

Let us now plot the mathematical function $y = x^2 - 3$ in the range x from -3 to $+3$. First, we need to define the vector x, then calculate the vector y using the above formula. Then we use the `plot` command as usual for the two vectors x and y. Finally, we use the other commands to display the title and axis label information on the graph.

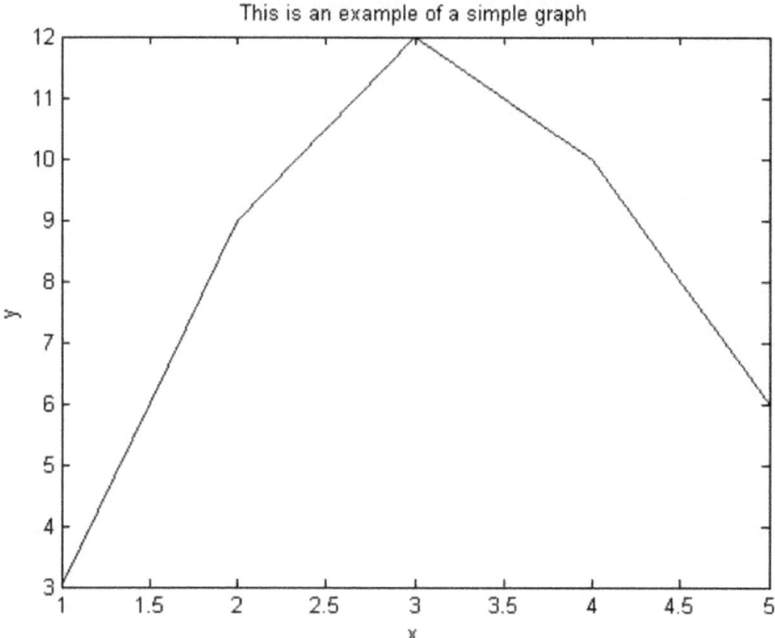

Figure 9.2: Title and axis information displayed on graph

The following are the needed commands. The resulting graph is shown in Figure 9.3.

```
>> x = [-3 -2 -1 0 1 2 3]

x =

    -3    -2    -1     0     1     2     3

>> y = x.^2 -3

y =

     6     1    -2    -3    -2     1     6
>> plot(x,y)
>> hold on
```

```
>> title('This is another example of the plot command')
>> xlabel('x-axis')
>> ylabel('y-axis')
```

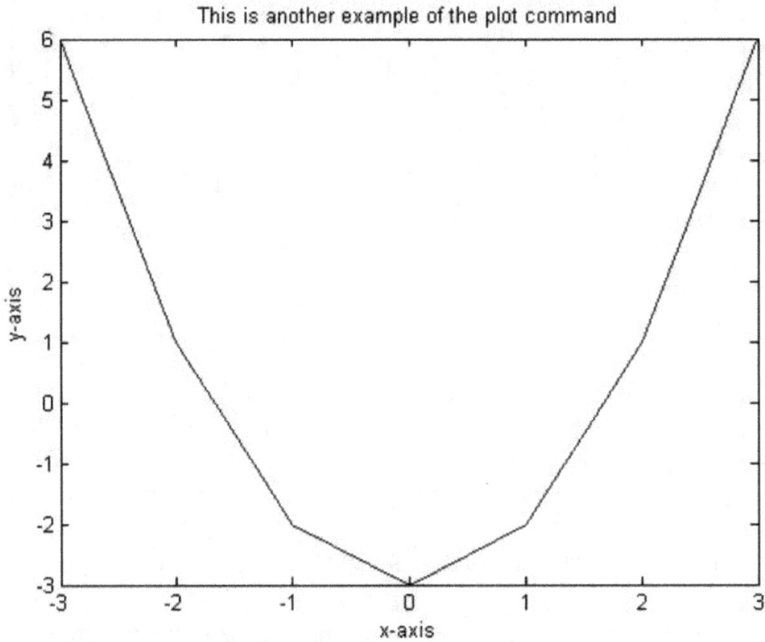

Figure 9.3: Example of plotting a mathematical function

It is seen from the above graphs that they are plotted with solid lines. The line style may be changed from solid to dotted, dashed, or dash-dot lines by changing the parameters in the plot command. Use the command `plot(x,y,':')` to get a dotted line, the command `plot(x,y,'--')` to get a dashed line, the command `plot(x,y,'-.')` to get a dash-dot line, and the default command `plot(x,y,'-')` to get a solid line.

In addition, the color of the plotted line may be changed. Use the command `plot(x,y,'b')` to get a blue line. Replace the b with g for green, r for red, c for cyan, m for magnetta, y for yellow, k for black, and w for white. Consult the MATLAB manual for a full list of the color options. Note that colors are not displayed in this book.

In addition to changing the line style and color, we can include symbols at the plotted points. For example, we can use the circle or cross symbols to denote the plotted points. Use the command `plot(x,y,'o')` to get a circle symbol. Replace the o with x to get a cross symbol, + for a plus sign symbol, * for a star symbol, s for a square symbol, d for a diamond symbol, and . for a point symbol. There are other commands for other symbols like different variations of triangles, etc. Again, consult the MATLAB manual for the full list of available symbols.

The above options for line style, color, and symbols can be combined in the same command. For example use the command `plot(x,y,'rx:')` to get a red dotted line with cross symbols. Here is the previous example with a red dotted line with cross symbols but without the title and label information (note that no colors appear in this book). See Figure 9.4 for the resulting graph.

```
>> plot(x,y,'rx:')
```

Next, consider the following example where we show another use of the `plot` command. In this example, we plot two curves on the same graph. We use the modified `plot` command as `plot(x,y,x,z)` to plot the two curves. MATLAB plots the two mathematical functions y and z (on the y-axis) as a function of x (on the x-axis). We also use the MATLAB command `grid` to show a grid on the plot. Here are the needed commands[16]. The resulting graph is shown in Figure 9.5.

```
>> x = 0:pi/20:2*pi;
>> y = sin(x);
>> z = cos(x);
```

---

[16] These commands may be entered individually on the command prompt or stored as a script in a script file. See Chapter 4 for details about scripts and script files.

```
>> plot(x,y,'-',x,z,'--')
>> hold on;
>> grid on;
>> title('Sine and Cosine Functions')
>> xlabel('x-axis')
>> ylabel('sine and cosine')
>> legend('sinx','cosx')
```

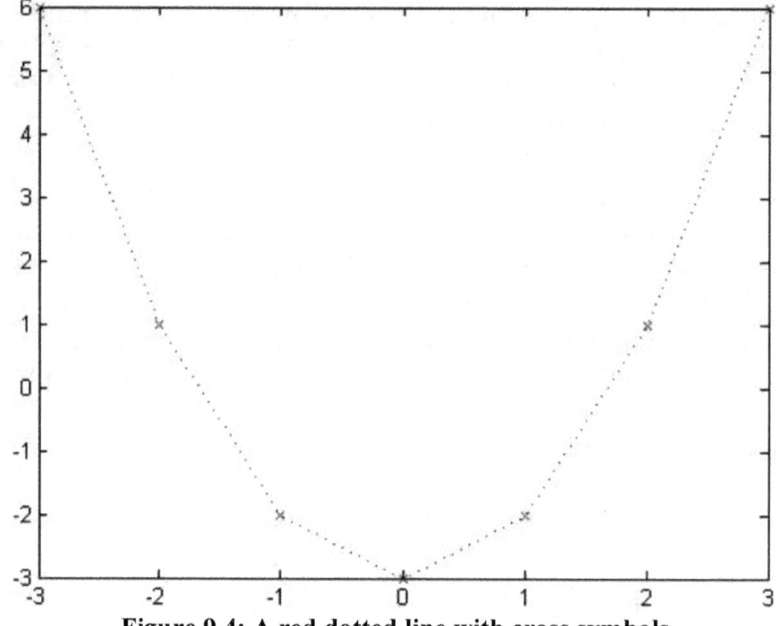

Figure 9.4: A red dotted line with cross symbols

Note in the above example that we used semicolons to suppress the output. Also, note the use of the MATLAB command `legend` to get a legend displayed at the top right corner of the graph. We can further use several `axis` commands to customize the appearance of the two axes including the tick marks, but this is beyond the scope of this book. Consult the MATLAB manual for further details about these commands.

There are some interactive plotting tools available in MATLAB to further customize the resulting graph. These tools are available from the menus of the graph window – the window where the graph appears.

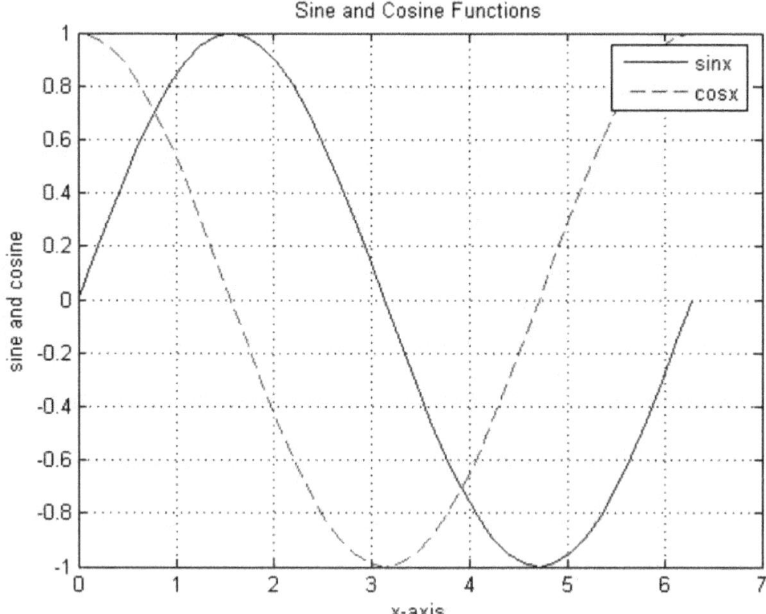

Figure 9.5: Example of two curves on the same graph

There are other specialized plots[17] that can be drawn automatically by MATLAB. For example, there are numerous plots that are used in statistics that can be generated by MATLAB. In particular, a pie chart can be generated by using the MATLAB command `pie`, a histogram can be generated by using the MATLAB command `hist`, and a rose diagram can be generated by using the MATLAB command `rose`. In addition, graphs of curves with polar coordinates can be generated by using the MATLAB command `polar`.

---

[17] Several logarithmic plots can also be generated by using the MATLAB command `loglog`. Semi-logarithmic plots can also be generated by using the MATLAB commands `semilogx` or `semilogy` but their use is beyond the scope of this book.

Our final two-dimensional graph will feature the use of the MATLAB command `subplot`. The use of this command will enable you to present several plot diagrams on the same graph. In this example, we use four instances of the `subplot` command to show four diagrams on the same plot. We will plot four mathematical functions – each on its own diagram – but the four diagrams will appear on the same graph. We can do this by playing with the parameters of the `subplot` command. Note that the diagrams are shown without title or axis labels in order to emphasize the use of the `subplot` command. Here are the necessary MATLAB commands[18] used with semicolons to suppress the output:

```
>> x = [ 1 2 3 4 5 6 7 8 9 10];
>> y = x.^2;
>> z = sqrt(x);
>> w = 2*x - 3;
>> v = x.^3;
>> subplot(2,2,1)
>> hold on;
>> plot(x,y)
>> subplot(2,2,2)
>> plot(x,z)
>> subplot(2,2,3)
>> plot(x,w)
>> subplot(2,2,4)
>> plot(x,v)
```

Note the use of the three parameters of the `subplot` command. The first two parameters `(2,2)` indicate that the graph area should be divided into four quadrants with each row and column comprising of two sub-areas for plots. The third parameter indicates in which sub-area the next plot will appear. Note also that each `subplot` command is followed by a corresponding `plot`

---

[18] These commands may be entered individually on the command prompt or stored in a script file.

command. Actually, each `subplot` command reserves a specific area for the plot while the subsequent corresponding `plot` command does the actual plotting. The resulting graph is shown in Figure 9.6. You can control the number and arrangement of the diagrams that are displayed by controlling the three parameters of the `subplot` command.

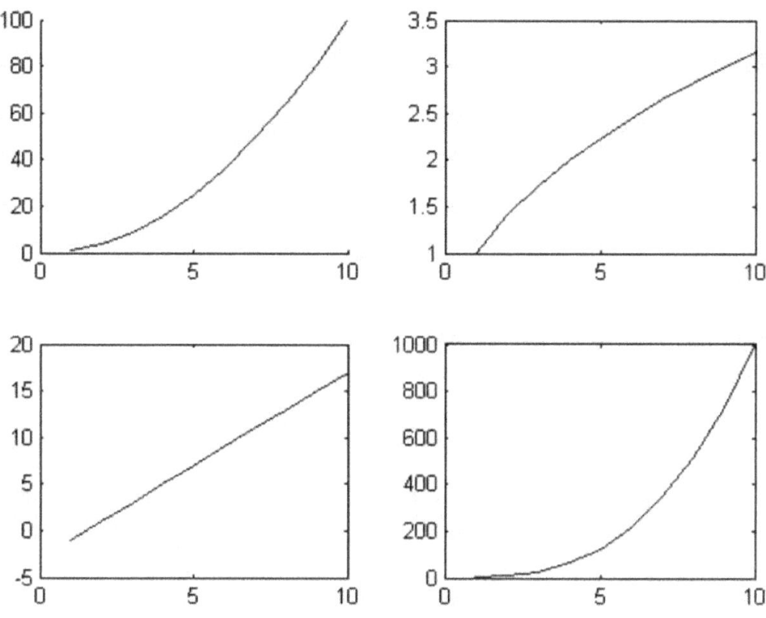

Figure 9.6: Use of the MATLAB command subplot

Next, we will discuss how to generate three-dimensional graphs with MATLAB. There are mainly four commands that can be used in MATLAB for plotting three-dimensional graphs. The first command is `plot3` which is used to plot curves in space. It requires three vectors as input. Here is an example of this command followed by the resulting graph as shown in Figure 9.7.

```
>> x = [1 2 3 4 5 6 7 8 9 10]
x =
```

```
         1         2         3         4         5         6         7
8        9        10

>> y = [1 2 3 4 5 6 7 8 9 10]

y =

         1         2         3         4         5         6         7
8        9        10

>> z = sin(x).*cos(y)

z =

  Columns 1 through 6

     0.4546    -0.3784    -0.1397     0.4947     -
0.2720    -0.2683

  Columns 7 through 10

     0.4953    -0.1440    -0.3755     0.4565

>> plot3(x,y,z)
>> hold on;
>> grid on;
```

It should be noted that the title, xlabel, and ylabel commands can also be used with three-dimensional graphs. For the z-axis, the MATLAB command zlabel is also used in a way similar to the xlabel and ylabel commands.

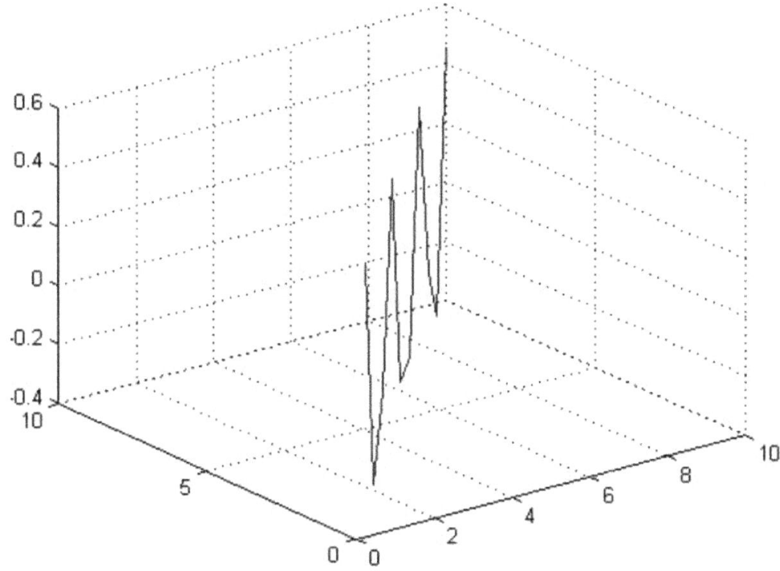

Figure 9.7: Example of using the plot3 command

The second command in MATLAB for three dimensional plots is mesh. The use of the mesh command is used to generate mesh surfaces. The simplest use of this command is with a matrix. It assumes that the matrix elements are values of a grid spanning the matrix dimensions. The values of the elements of the matrix are taken automatically along the z-axis while the numbers of rows and columns are taken along the x- and y-axes. Here is an example followed by the resulting graph as shown in Figure 9.8.

```
>> A = [10 14 20 37 5 17 11 ; 12 20 40 20 11
5 14 ; 30 51 12 17 20 30 2 ; 24 34 56 10 14
5 40 ; 34 12 33 12 26 10 15 ; 12 45 13 23 35
10 7 ; 10 20 13 34 32 10 7]

A =

        10        14        20        37         5        17        11
        12        20        40        20        11         5        14
        30        51        12        17        20        30         2
```

24	34	56	10	14	5	40
34	12	33	12	26	10	15
12	45	13	23	35	10	7
10	20	13	34	32	10	7

```
>> mesh(A)
```

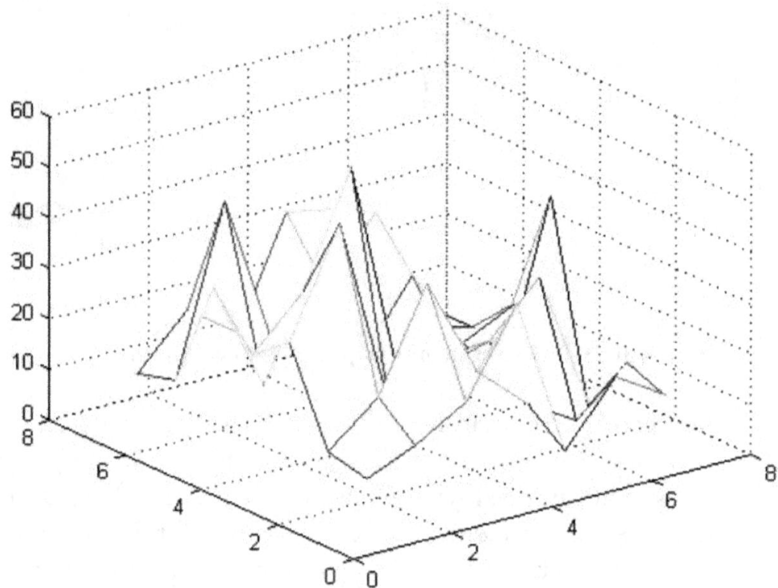

Figure 9.8: Use of the mesh command

If a matrix or a grid is not available, then a grid can be generated using the MATLAB command `meshgrid` but this is beyond the scope of this book. Again, the reader should use the title and axis information commands to display information on the graph but these are not shown in this example.

The third MATLAB command for three-dimensional plots is the `surf` command. Its use is similar to the `mesh` command. Here is an example of its use with the same matrix A of the previous example followed by the resulting graph as shown in Figure 9.9.

```
>> surf(A)
```

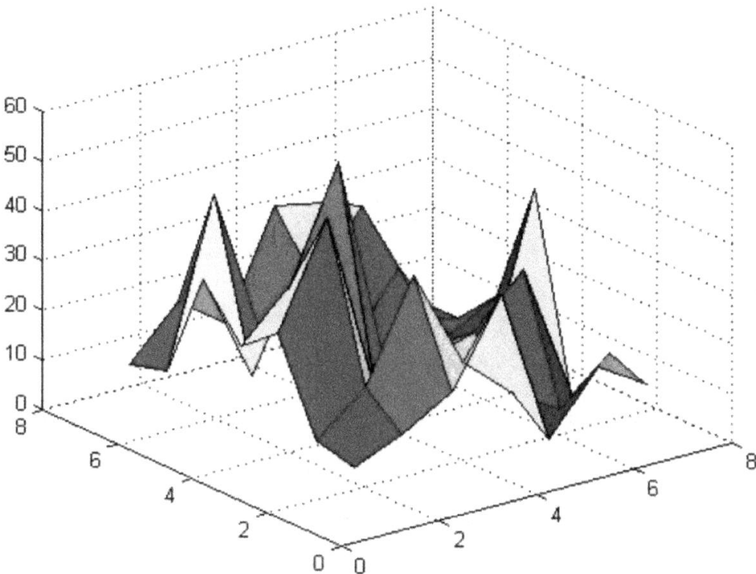

Figure 9.9: Use of the surf command

The above three-dimensional plot is usually shown in color in MATLAB but we are not able to display colors in this book. Again, the MATLAB command `meshgrid` may be used to generate a grid for the three-dimensional plot if one is not available. Again, the reader should use the title and axis information commands to display these types of information on the plot above but these are not used in this example.

The fourth command for three-dimensional plots is the MATLAB command `contour`. The use of this command produces contour plots of three-dimensional surfaces. The use of this command is similar to the `mesh` and `surf` commands. Here is an example of its use with the same matrix A that was used previously followed by the resulting graph as shown in Figure 9.10.

```
>> contour(A)
```

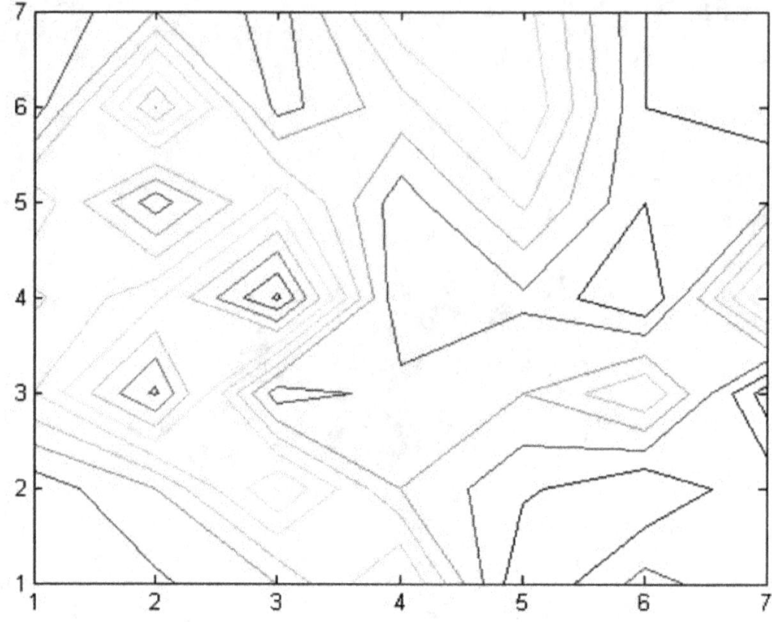

Figure 9.10: Use of the contour command

The above contour plot is usually shown in color in MATLAB but we are not able to display colors in this book. Again, the MATLAB command `meshgrid` may be used to generate a grid for the contour plot if one is not available. The reader can add the contour heights to the graph using the MATLAB command `clabel`.

There are some variations of the `mesh` and `surf` commands in MATLAB. For example, the `meshc` and `surfc` commands produce the same mesh and surface plots as the commands `mesh` and `surf` but with the contour plot appearing underneath the surface. Here is an example of the use of the `surfc` command with the same matrix A used previously followed by the resulting graph as shown in Figure 9.11.

```
>> surfc(A)
```

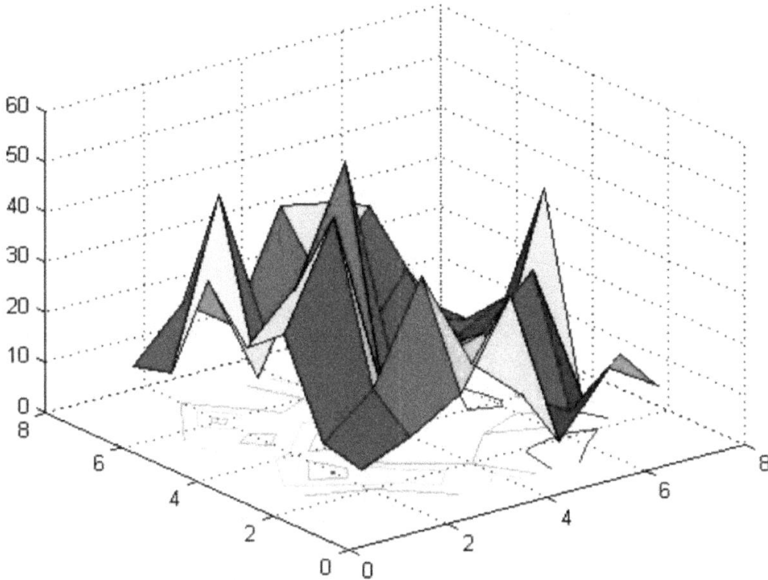

Figure 9.11: Use of the surfc command

Another variation is the `meshz` command which produces the same three-dimensional plot as the `mesh` command but with a zero plane underneath the surface. The use of this command is very simple and is not shown here.

MATLAB Guide to Fibonacci Numbers and the Golden Ratio
A Simplified Approach, for Students and Scientists and Engineers

Part II

Fibonacci Numbers
and
The Golden Ratio

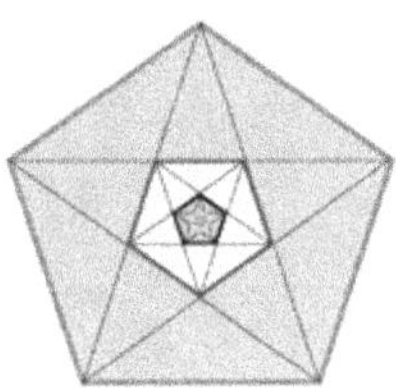

MATLAB Guide to Fibonacci Numbers and the Golden Ratio

A Simplified Approach, for Students and Scientists and Engineers

6. Fibonacci Numbers

The Fibonacci sequence is a sequence of numbers where each term of the sequence is obtained by adding the previous two terms. In addition, this special sequence starts with the numbers 1 and 1. Thus, the whole Fibonacci sequence can be generated starting with these two numbers and the above special rule. The numbers in this sequence are called Fibonacci numbers. These numbers were first obtained by the Italian mathematician Leonardo of Pisa (also called Leonardo Fibonacci) in the thirteenth century.

Next, we will generate the first few terms of the Fibonacci sequence. We will start with 1 and 1[19]. Adding these two terms brings us to 2 which is the third Fibonacci number. Then we add 2 and 1 to obtain 3 which is the fourth Fibonacci number. Next, we add 3 and 2 to obtain 5 which is the fifth Fibonacci number. Then, we add 5 and 3 to obtain 8 which is the sixth Fibonacci number. This is followed by adding 8 and 5 to obtain 13 which is the seventh Fibonacci number. This process is illustrated interactively in MATLAB as follows:

```
>> 1 + 1

ans =

     2

>> 2 + 1

ans =

     3
```

---

[19] The Fibonacci sequence can also be started with the numbers 0 and 1 instead of 1 and 1. Similar results will be obtained in this case.

```
>> 3 + 2

ans =

     5

>> 5 + 3

ans =

     8

>> 8 + 5

ans =

    13

>> 13 + 8

ans =

    21

>> 21 + 13

ans =

    34

>> 34 + 21

ans =

    55

>> 55 + 34
```

```
ans =

    89

>> 89 + 55

ans =

    144

>> 144 + 89

ans =

    233
```

The sequence of numbers obtained this way is 1, 1, 2, 3, 5, 8, 13, 21, 34, 55, 89, 144, 233, ... is called the Fibonacci sequence. More terms can be generated but the above thirteen terms will suffice for now. Obviously, the Fibonacci sequence goes all the way to infinity as the number of terms increase to infinity. The mathematics of the Fibonacci sequence will be discussed later in this book.

We can write a MATLAB function to readily generate the Fibonacci sequence up until the n-th term. This MATLAB function will be called `fibonacci(n)`. The coding for this function is illustrated below:

```
function f=fibonacci(n)
% This function generates the first n
% Fibonacci numbers
f = zeros(n,1);
f(1)=1;
f(2)=2;
for k = 3:n
    f(k) = f(k-1) + f(k-2);
end
```

The above MATLAB function can be executed many times; each time with a different value for n, depending on how many terms are needed in the Fibonacci sequence. For example, the first seven Fibonacci numbers are generated by substituting 7 for n in the function `fibonacci(n)` as follows;

```
>> fibonacci(7)

ans =

     1
     2
     3
     5
     8
    13
    21
```

The function `fibonacci(n)` can be executed again and again as many times as necessary. For example, the following execution of this function generates the first fifteen Fibonacci numbers;

```
>> fibonacci(15)

ans =

     1
     2
     3
     5
     8
    13
    21
    34
    55
    89
```

```
144
233
377
610
987
```

Similarly, the first thirty terms of the Fibonacci sequence are generated as follows:

```
>> fibonacci(30)

ans =

                1
                2
                3
                5
                8
               13
               21
               34
               55
               89
              144
              233
              377
              610
              987
             1597
             2584
             4181
             6765
            10946
            17711
            28657
            46368
            75025
```

```
      121393
      196418
      317811
      514229
      832040
     1346269
```

Thus, we see that the Fibonacci numbers increase exponentially as the value of n increases. We will illustrate this property of the Fibonacci sequence graphically next. But first, let us investigate how to display only the desired Fibonacci number in MATLAB. To accomplish this, we store the Fibonacci sequence in a variable, e.g. y while suppressing the output, then we invoke y(n) for the desired nth Fibonacci number. Here is an example to accomplish this trick in MATLAB.

```
>> y = fibonacci(15);
>> y(15)

ans =

  987
```

Thus, it is seen from the above example that the 15th Fibonacci number is 987. The above trick is especially helpful if we do not want to display the complete Fibonacci sequence of numbers. Another alternative is to revise the MATLAB code for the function fibonacci(n) slightly to display only the desired outcome.

Let us write the rule for generating the Fibonacci sequence using mathematical notation. If we denote the nth term of the Fibonacci sequence by F(n), then F(n+1) would be the next consecutive Fibonacci number. The previous two Fibonacci numbers would be denoted by F(n-1) and F(n-2). Consequently, the nth Fibonacci number can be generated by the following rule:

$$F(n) = F(n-1) + F(n-2)$$

$$F(1) = 1$$

$$F(2) = 1$$

$$n = 3, 4, 5, \ldots$$

To illustrate using the above mathematical notation, let us show an example with n = 5. This would generate the fifth Fibonacci number. Substituting in the first equation above n = 5, we obtain:

$$F(5) = F(4) + F(3)$$

To compute both F(4) and F(3), we substitute each one in the same equation again (this equation is called a recursive relation). This, we obtain;

$$F(4) = F(3) + F(2)$$

$$F(3) = F(2) + F(1)$$

Using the initial conditions that F(1) = 1 and F(2) = 1, and substituting in the above relations, we obtain:

$$F(3) = F(2) + F(1) = 1 + 1 = 2$$

$$F(4) = F(3) + F(2) = 2 + 1 = 3$$

$$F(5) = F(4) + F(3) = 3 + 2 = 5$$

Thus, the fifth Fibonacci number is 5.

Let us plot a graph showing the relationship between the number of terms and the corresponding Fibonacci numbers. In this first plot, let us use only the first ten terms of the Fibonacci sequence. First, let us generate the x-axis of the plot as follows:

```
>> x = 1:10

x =

     1     2     3     4     5     6     7
     8     9    10
```

The next step would be to generate the y-axis which would be the first ten Fibonacci numbers. These are generated and stored in the variable y as follows:

```
>> y = fibonacci(10)

y =

     1
     2
     3
     5
     8
    13
    21
    34
    55
    89
```

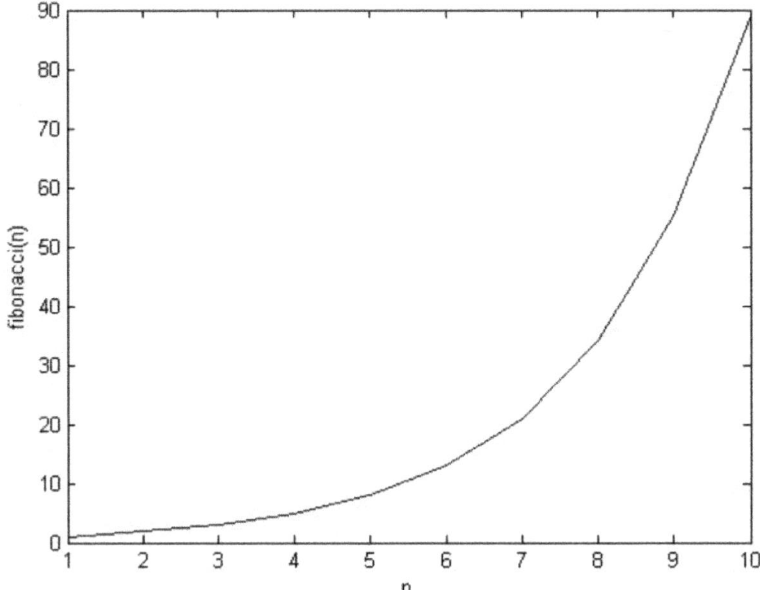

Figure 6.1: Graph of the first ten Fibonacci numbers

The final step would be to use the MATLAB command plot in order to generated the required graph. In addition, we use the MATLAB commands xlabel and ylabel to label the two axes, respectively. These commands are executed as shown below and the graph is shown in Figure 6.1.

```
>> plot(x,y)
>> hold on
>> xlabel('n')
>> ylabel('fibonacci(n)')
```

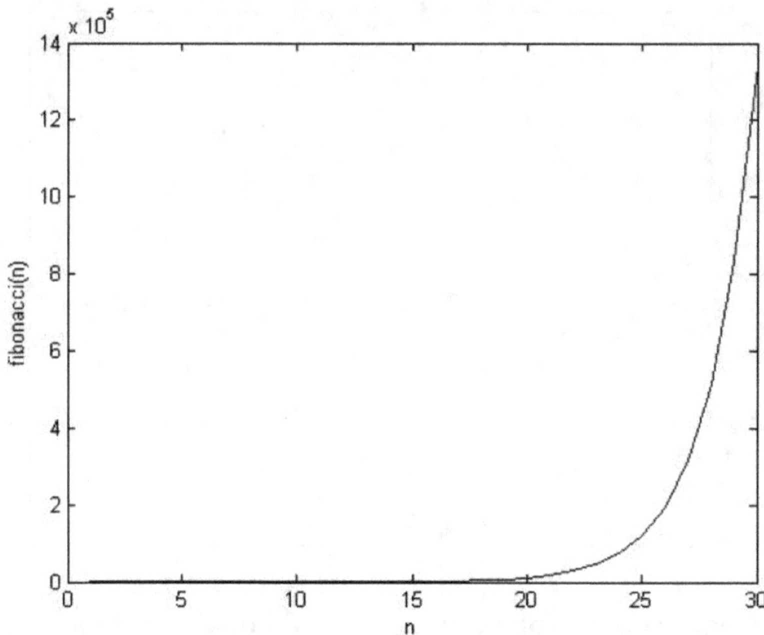

Figure 6.2: Graph of the first thirty Fibonacci numbers

Similarly, the following commands will generate and plot the first 30 Fibonacci numbers. The plot is generated after the commands are executed sequentially and is shown in Figure 6.2. Each command is followed by a semi-colon in order to suppress the listed output.

```
>> x = 1:30;
>> y = fibonacci(30);
>> plot(x,y)
>> hold on;
>> xlabel('n')
>> ylabel('fibonacci(n)')
```

It is seen from the above graph that Fibonacci numbers increase exponentially as the value of n increases. In the next chapter, we will investigate the Fibonacci sequence in more details and study its exact property that makes this sequence special – this would be its relationship to what is called the Golden Ratio.

Using the mathematical notation of this chapter, the two graphs in Figures 6.1 and 6.2 are called *F(n) vs. n graphs*. Next, we will study the graphs obtained when plotting the logarithms of Fibonacci numbers.

First, let us use the MATLAB function `log` to calculate the logarithms[20] of the first ten Fibonacci numbers as follows:

```
>> log(fibonacci(10))

ans =

        0
   0.6931
   1.0986
   1.6094
   2.0794
   2.5649
   3.0445
   3.5264
   4.0073
   4.4886
```

We need now to plot the above logarithms on the y-axis while keeping the number of terms n on the x-axis. The type of plot obtained is called a semi-log graph. We will perform the following operations (while suppressing the output) in order to accomplish this. The resulting graph is shown in Figure 6.3. This graph is called a *log(F(n)) vs. n graph*. The MATLAB commands are shown below.

```
>> x = 1:10;
```

---

[20] The logarithms calculated here are the natural logarithms, which are logarithms to the base e , where e = 2.71828....

```
>> y = log(fibonacci(10));
>> plot(x,y)
>> hold on
>> xlabel('n')
>> ylabel('log(F(n))')
```

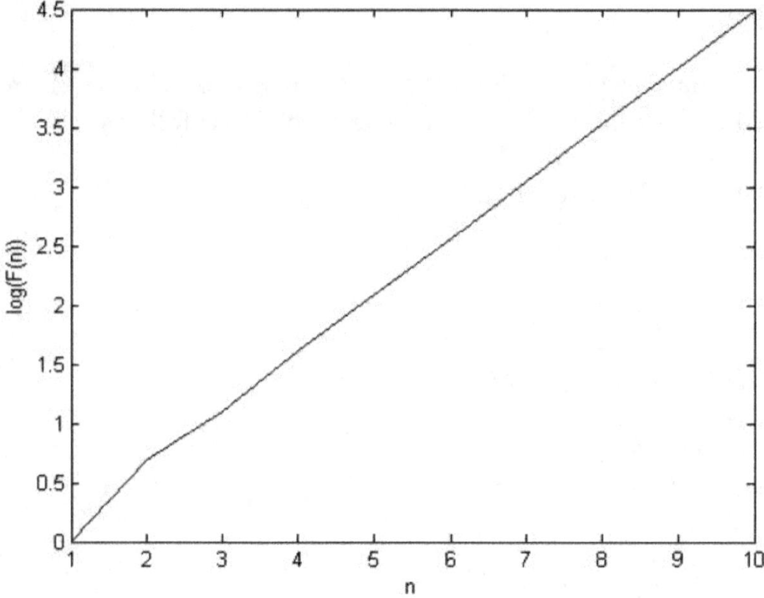

Figure 6.3: Semi-log graph of the first 10 Fibonacci numbers

It is seen from Figure 6.3 that the graph is almost a straight line. Let us calculate the slope[21] of this line. Let us perform this calculation by taking the fifth and tenth Fibonacci numbers and calculating the slope of the line segment between them. This is implemented in MATLAB as follows:

```
>> m = (y(10)-y(5))/(x(10)-x(5))

m =
```

---

[21] The slope m of a line segment between two points (x_1, y_1) and (x_2, y_2) is calculated using the formula $m = (y_2 - y_1)/(x_2 - x_1)$.

```
0.4818
```

Thus the slope of the straight line is 0.4818. In order to find the angle, we calculate the inverse tangent value of this number. This is obtained as $\tan^{-1}(0.4818) = 25.7°$.

Let us perform the above operations again but using the first 30 Fibonacci numbers. Here are the MATLAB commands to generate the plot while the graph is shown in Figure 6.4.

```
>> x = 1:30;
>> y = log(fibonacci(30));
>> plot(x,y)
>> hold on
>> xlabel('n')
>> ylabel('log(F(n))')
```

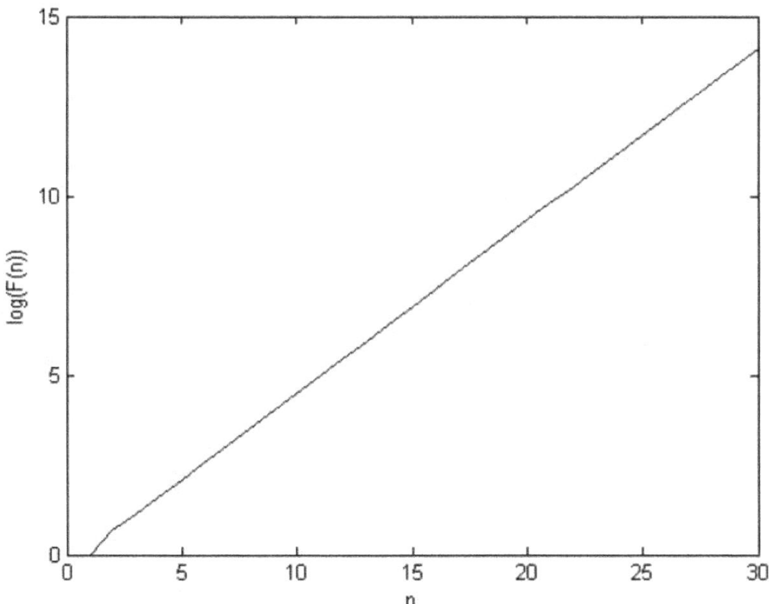

Figure 6.4: Semi-log graph of the first 30 Fibonacci numbers

As expected, the semi-log graph shows almost a straight line. Let us now calculate the slope of this line between the 10th and the 30th Fibonacci numbers. This operation is implemented in MATLAB as follows:

```
>> m = (y(30)-y(10))/(x(30)-x(10))

m =

    0.4812
```

As expected, the slope of the line is almost as before. The angle of the line is calculated as $\tan^{-1}(0.4812) = 25.7°$, exactly as before. An alternative method to plot the logarithms of Fibonacci numbers is to use the MATLAB command `semilogy` in association with the MATLAB function `fibonacci(n)`. The graph obtained in this way will also be an almost straight line but will be slightly different from the graphs obtained here.

Fibonacci numbers are closely related to exponential growth in nature. Botanists noticed that there are many plants that tend to have a Fibonacci number for the leaves and petals. Fibonacci numbers also appear prominently in the family tree of rabbits. Actually, Leonardo Fibonacci first noticed these numbers when he studied the problem of breeding of rabbits under ideal circumstances. Let us elaborate on this problem and how it is related to Fibonacci numbers. See the books in references [48-56] and the web links [57-67].

Consider a hypothetical situation of rabbits breeding as follows. Start with two rabbits who after a time produce two new rabbits. After a certain time, these four rabbits produce four new rabbits. Thus, we end up with eight rabbits. In the next cycle of breeding, another eight rabbits are produced to make the total 16 rabbits and so on. Thus, we get the exponential sequence 2, 4, 8, 16, 32, ... which is just a sequence of the powers of 2, i.e. 2^1, 2^2, 2^3, 2^4,

2^5, ... This sequence follows the law of exponential growth. See the books in references [48-56] and the web links [57-67].

However, the real situation is different. When we have four rabbits, the newborn rabbits are too young to produce new rabbits right away. They have to skip a cycle before they start producing more rabbits. In this way, the number of rabbits accumulating follows the Fibonacci sequence 1, 1, 2, 3, 5, 8, 13, 21, ... and so on. This is how Leonardo Fibonacci first deduced this interesting series by considering how rabbits breed naturally. This is explained in detail below. See the books in references [48-56] and the web links [57-67].

Let us start with one rabbit. After one cycle, this rabbit produces a new rabbit – the total becomes two. After the second cycle, the old rabbit produces another new rabbit but the little rabbit skips this cycle and does not produce any new rabbits. The total now becomes three rabbits. After another cycle, the two old rabbits produce two new rabbits but the little one skips this cycle and does not produce. Thus the total now becomes five rabbits following the Fibonacci sequence precisely. Thus, we obtain the numbers 1, 2, 3, 5, 8, 13, 21, ... for the total number of rabbits. This is exactly what Leonardo Fibonacci originally observed and is what started the interesting topic of Fibonacci numbers. See the books in references [48-56] and the web links [57-67].

Fibonacci numbers appear in nature often enough to show that they reflect some naturally occurring patterns. These patterns can be spotted by studying how certain plants grow. For example, if you look at an array of seeds in the center of a sunflower, you will notice what looks like spiral patterns curving left and right. If you count these spirals, you will be surprised to obtain a Fibonacci number. If you divide the spirals into those pointed left and right, you will get two consecutive Fibonacci numbers. You can find such spiral patterns in pinecones, pineapples, and cauliflower. See the books in references [48-56] and the web links [57-67].

Other examples where Fibonacci numbers occur naturally are in plant growth and branches of plants. For example, if you count the number of petals on a flower, you will be surprised to find the total to be a Fibonacci number. Lilies and irises have three petals while buttercups and wild roses have five. Delphiniums have eight petals and so on. Notice that 3, 5, and 8 are Fibonacci numbers. See the books in references [48-56] and the web links [57-67].

Fibonacci numbers also occur naturally in honeybee colonies. In these colonies, there is a queen, a few drones, and a lot of workers. The female bees all have two parents — a drone and a queen. However, drones have one parent each. Therefore, Fibonacci numbers appear in a drone's family tree — each drone has one parent, two grandparents, three great-grandparents, and so on. See the books in references [48-56] and the web links [57-67].

A final example is that Fibonacci numbers appear in the human body. You have one nose, two eyes, three segments to each limb, and five fingers on each hand. These are all Fibonacci numbers. Furthermore, DNA molecules follow the Fibonacci sequence also measuring 34 angstroms long and 21 angstroms wide for each full cycle of the double helix. Of course, the numbers 34 and 21 are surprisingly Fibonacci numbers. See the books in references [48-56] and the web links [57-67].

7. The Golden Ratio

In this chapter, we will continue our study of the Fibonacci sequence and illustrate its precise relationship to what is called the Golden Ratio. Let us calculate the ratio of each two consecutive numbers in the Fibonacci sequence.

Here are the first few terms of the Fibonacci sequence:

1, 1, 2, 3, 5, 8, 13, 21, 34, 55, 89, 144, ...

Let us start by calculating the ratio of the first two terms. The first two terms are 1 and 1; their ratio is $1/1 = 1$. The next two consecutive terms are 1 and 2; their ratio is $2/1 = 2$. Next are the numbers 2 and 3; their ratio is $3/2 = 1.5$. The ratio of 3 and 5 is $5/3 = 1.6667$. Next are the numbers 5 and 8 whose ratio is $8/5 = 1.6$. This followed by the numbers 8 and 13 whose ratio is $13/8 = 1.625$. Next are 13 and 21 whose ratio is $21/13 = 1.6154$. This is followed by the numbers 21 ad 34; their ratio is $34/21 = 1.6190$. Next are the numbers 34 and 55 whose ratio is $55/34 = 1.6176$. The ratio of the next two numbers in the sequence is $89/55 = 1.6182$. Finally, the last two terms in the sequence above have the ratio $144/89 = 1.6180$.

As can be seen from the above calculations, the ratio of any two consecutive Fibonacci numbers seems to approach the number 1.6180 as the numbers increase. In order to ascertain this observation, the above calculations are easily implemented using MATLAB. First, let us perform the above calculations again using MATLAB interactively as follows:

```
>> 1/1

ans =

    1
```

```
>> 2/1

ans =

     2

>> 3/2

ans =

    1.5000

>> 5/3

ans =

    1.6667

>> 8/5

ans =

    1.6000

>> 13/8

ans =

    1.6250

>> 21/13

ans =

    1.6154

>> 34/21
```

```
ans =

    1.6190

>> 55/34

ans =

    1.6176

>> 89/55

ans =

    1.6182

>> 144/89

ans =

    1.6180

>> 233/144

ans =

    1.6181

>> 377/233

ans =

    1.6180

>> 610/377

ans =
```

```
   1.6180

>> 987/610

ans =

   1.6180
```

It is clearly seen from the above MATLAB output that the sought ratio approaches the magical number 1.6180. Next, a MATLAB function is shown which calculates this ratio as above and plots the graph of the ratio as the number of terms increases. Here is the code of the MATLAB function `ratio(n)` which calculates this limiting ratio:

```
function r=ratio(n)
% This function calculates the ratio of two
%  consecutive Fibonacci numbers
f = zeros(n,1);
r = zeros(n,1);
f(1)=1;
f(2)=2;
for k = 3:n
    f(k)  =  f(k-1)  +  f(k-2);
    r(k)  =  f(k)/f(k-1);
end
```

Let us execute the above functions for the first fifteen terms of the Fibonacci sequence. Here is the output of this command:

```
>> ratio(15)

ans =

        0
        0
   1.5000
```

```
1.6667
1.6000
1.6250
1.6154
1.6190
1.6176
1.6182
1.6180
1.6181
1.6180
1.6180
1.6180
```

It is seen from the above output that the ratio of two consecutive Fibonacci numbers approaches the value 1.6180. To get more precision, we will use the format long command and execute the function again but with thirty terms as follows:

```
>> format long
>> ratio(30)

ans =

                              0
                              0
   1.50000000000000
   1.66666666666667
   1.60000000000000
   1.62500000000000
   1.61538461538462
   1.61904761904762
   1.61764705882353
   1.61818181818182
   1.61797752808989
   1.61805555555556
   1.61802575107296
   1.61803713527851
```

```
1.61803278688525
1.61803444782168
1.61803381340013
1.61803405572755
1.61803396316671
1.61803399852180
1.61803398501736
1.61803399017560
1.61803398820533
1.61803398895790
1.61803398867044
1.61803398878024
1.61803398873830
1.61803398875432
1.61803398874820
1.61803398875054
```

It is seen from the above results that the limiting ratio approaches the value 1.6180339887….. (accurate to eleven decimal digits). Let us plot a graph of this ratio with the following commands where the output is suppressed:

```
>> x = 1:30;
>> y = ratio(30);
>> plot(x,y)
>> hold on;
>> xlabel('n');
>> ylabel('F(n)/F(n-1)')
```

The resulting graph is shown in Figure 7.1. It is seen that the ratio F(n)/F(n-1) approaches the limiting value 1.61803.... . The graph shown in the figure is known as the *F(n)/F(n-1) vs. n* graph.

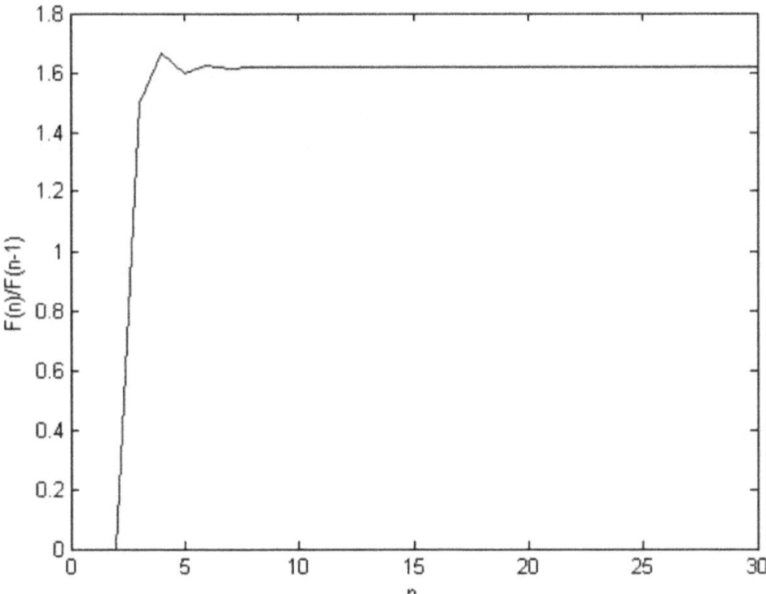

Figure 7.1: Graph of the limiting ratio of the Fibonacci sequence (shown for the first thirty terms of the sequence). The limiting ration clearly approaches the value 1.61803...

Next, we will investigate the number 1.61803... in more detail. This number is called the Golden Ratio or Golden Section or Divine Proportion. It will be seen that this number arises naturally from certain geometrical constructions. For example, if we divide a straight line into two segments according to a certain rule, we will obtain the Golden Ratio.

Consider a straight line AB which we would like to divide into two segments AC and CB as shown in Figure 7.2. The rule that we will use to obtain the point C is that the ratio of the length of the straight line to the length of the longer segment is equal to the ratio of the length of the longer segment to the length of the shorter segment. This ratios AB/AC and AC/CB should be equal. Thus, we have the following equation:

Figure 7.2: Dividing a straight line into two segments in deriving the Golden Ratio

$$\frac{AB}{AC} = \frac{AC}{CB}$$

However, we know that AB = AC + CB. Substituting this relation in the above equation, we obtain:

$$\frac{AC + CB}{AC} = \frac{AC}{CB}$$

Next, we will re-write the above equation as follows:

$$1 + \frac{CB}{AC} = \frac{AC}{CB}$$

Let the ratio AC/CB be denoted by the variable x, then the above equation may be re-written as follows:

$$1 + \frac{1}{x} = x$$

Multiplying both sides of the above equation by x, we obtain the following quadratic equation:

$$x + 1 = x^2$$

Next, we will write the above equation in the standard form of a quadratic equation as follows (by re-arranging the terms):

$$x^2 - x - 1 = 0$$

The above equation can be solved using the quadratic formula but we will use MATLAB to obtain the two solutions. In order to accomplish this, we use the MATLAB command `solve`[22] as follows:

```
>> solve('x^2 - x -1 = 0')

ans =

1/2*5^(1/2)+1/2
1/2-1/2*5^(1/2)
```

Thus, it is seen that the two solutions obtained are $\dfrac{1+\sqrt{5}}{2}$ and $\dfrac{1-\sqrt{5}}{2}$. Clearly, the second solution $\dfrac{1-\sqrt{5}}{2}$ is a negative number and should be discarded with regard to our example. Therefore, the positive solution $\dfrac{1+\sqrt{5}}{2}$ obtained is the one that is adopted.

Using MATLAB, we can now easily calculate the numerical value of the positive solution as follows:

```
>> (1+sqrt(5))/2
ans =

1.6180
```

---

[22] The `solve` command is part of the MATLAB Symbolic Math Toolbox. You need to have this toolbox installed along with MATLAB in order to use this command.

Thus, it is seen that the solution is exactly the Golden Ratio – the limiting ratio of the Fibonacci sequence. In order to ascertain this, we repeat the above calculation but with more precision as follows:

```
>> format long
>> (1+sqrt(5))/2

ans =

    1.61803398874989
```

Exactly, the same magical number is obtained which is the Golden Ratio. This number is usually denoted by the Greek symbol ϕ. Thus, we have $\phi = 1.618033987....$.

Going back to the two solutions of the quadratic equation, let us find the numerical value of the second solution (the negative solution that was discarded) which is $\dfrac{1-\sqrt{5}}{2}$. Using MATLAB, we find its numerical value as follows:

```
>> (1-sqrt(5))/2

ans =

    -0.6180
```

It is seen from the above calculation that the numerical value of the second solution is related to the value of the Golden Ratio. In order to find the exact relationship between the two solutions, let us repeat the above calculation but using more precision as follows:

```
>> format long
```

```
>> (1-sqrt(5))/2

ans =

  -0.61803398874989
```

It is seen from the above calculation that the second solution is equal to $1-\phi$. Thus the two solutions of the quadratic equation are ϕ and $1-\phi$. In order to check these that these two values are really the solutions, we substitute each one back into the quadratic equation as follows:

$$\phi^2 - \phi - 1 = 0$$

$$(1-\phi)^2 - (1-\phi) - 1 = 0$$

The above two equations are checked with MATLAB as follows:

```
>> phi = (1+sqrt(5))/2

phi =

    1.6180

>> phi^2 - phi - 1

ans =

     0

>> (1-phi)^2 - (1-phi) - 1
ans =

     0
```

Thus, it is ascertained that both values are solutions of the quadratic equation. We summarize what we have obtained thus far as follows:

$$\phi = \frac{1+\sqrt{5}}{2} = 1.61803398874989$$

$$\phi = \frac{1-\sqrt{5}}{2} = -0.61803398874989$$

In the next chapter, we will investigate certain properties of the Golden Ratio ϕ in more detail.

8. Properties of the Golden Ratio

In this chapter, we will study certain properties of the Golden Ratio in detail. In addition, we will study a nice formula that is used to compute Fibonacci numbers using the value of the Golden Ratio.

First, let us calculate the inverse of the Golden Ratio ϕ, i.e. let us calculate numerically the value of $\dfrac{1}{\phi}$. This computation is performed in MATLAB as follows:

```
>> phi = (1+sqrt(5))/2

phi =

    1.6180

>> 1/phi

ans =

    0.6180
```

It is seen from the above computation that the value of $\dfrac{1}{\phi}$ is surprisingly equal to $\phi-1$ which is 0.6180. Thus, we obtain the following formula for ϕ:

$$\frac{1}{\phi} = \phi - 1$$

In order to check the validity of the above formula, we multiply both sides by ϕ to get our familiar quadratic equation $\phi^2 - \phi - 1 = 0$.

Next, let us calculate the square of the Golden Ratio, i.e. the value of ϕ^2. This operation is performed in MATLAB as follows:

```
>> phi = (1+sqrt(5))/2

phi =

    1.6180

>> phi^2

ans =

    2.6180
```

It is seen that another surprising result is obtained. The value of ϕ^2 is seen to be equal to $\phi + 1$ which is 2.6180. Thus, we have the following relation:

$$\phi^2 = \phi + 1$$

Again we obtain the same familiar quadratic equation $\phi^2 - \phi - 1 = 0$.

We have found out that the Golden Ratio has some interesting properties: its reciprocal is equal to $\phi - 1$ and its square is equal to $\phi + 1$. Let us see what other interesting properties this number has.

Let us now study the powers of ϕ, i.e. let us study the terms ϕ^2, ϕ^3, ϕ^4, ... and so on. These powers are given by the following set of equations:

$$\phi^2 = \phi = 1$$

$$\phi^3 = \phi^2 + \phi$$

$$\phi^4 = \phi^3 + \phi^2$$

$$\cdots\cdots\cdots\cdots$$

$$\phi^{n+2} = \phi^{n+1} + \phi^n \quad , \quad n = 0,1,2,3,\ldots$$

We have already checked the second power of ϕ using MATLAB before. Let us now check the third and fourth power of ϕ using MATLAB in order to verify the above equations. Here is the MATLAB output to check the equation for ϕ^3:

```
>> phi = (1+sqrt(5))/2

phi =

    1.6180

>> phi^3

ans =

    4.2361

>> phi^2 + phi

ans =

    4.2361
```

Now, let us show the MATLAB output to check the equation for ϕ^4 as follows:

```
>> phi = (1+sqrt(5))/2

phi =

    1.6180

>> phi^4

ans =

    6.8541

>> phi^3 + phi^2

ans =

    6.8541
```

Thus, the above equations for the powers of ϕ are verified. There is another set of equations for the powers of ϕ that is also interesting. This new set of equations is written as follows:

$$\phi^2 = \phi + 1$$

$$\phi^3 = 2\phi + 1$$

$$\phi^4 = 3\phi + 2$$

$$\phi^5 = 5\phi + 3$$

$$\phi^6 = 8\phi + 5$$

$$\phi^7 = 13\phi + 8$$

We can see that a pattern emerges for the expressions of the powers of ϕ shown above. It is clear that the numbers appearing in the expressions on the right-hand-side of the equations are Fibonacci numbers. Thus, the above equations serve to strengthen the relationship between Fibonacci numbers and the Golden Ratio. Let us now check the equations for ϕ^3 and ϕ^4 shown above using MATLAB. First, the expression for ϕ^3 is verified as follows:

```
>> phi = (1+sqrt(5))/2

phi =

    1.6180

>> phi^3

ans =

    4.2361

>> 2*phi + 1

ans =

    4.2361
```

Next, the expression for ϕ^4 is also verified as follows:

```
>> phi = (1+sqrt(5))/2

phi =

    1.6180
```

```
>> phi^4

ans =

    6.8541

>> 3*phi +2

ans =

    6.8541
```

Next, we will derive the expressions for the previous two equations for ϕ^3 and ϕ^4 mathematically as follows. Let us start with the expression $\phi^3 = \phi^2 + \phi$ and substitute $\phi + 1$ for ϕ^2 to obtain $\phi^3 = (\phi + 1) + \phi = 2\phi + 1$. Similarly, we start with the expression $\phi^4 = \phi^3 + \phi^2$ and substitute $2\phi + 1$ for ϕ^3, and $\phi + 1$ for ϕ^2 to obtain $\phi^4 = (2\phi + 1) + (\phi + 1) = 3\phi + 2$. We can continue in a similar fashion to prove the other expressions shown above and generate the Fibonacci numbers in this way.

Next, let us study the inverse powers of ϕ, i.e. let us study the expressions for $\dfrac{1}{\phi}$, $\dfrac{1}{\phi^2}$, $\dfrac{1}{\phi^3}$, $\dfrac{1}{\phi^4}$, ... and so on. The following expressions will be shown to be true for the inverse powers of the Golden Ratio:

$$1 = \frac{1}{\phi} + \frac{1}{\phi^2}$$

$$\frac{1}{\phi} = \frac{1}{\phi^2} + \frac{1}{\phi^3}$$

$$\frac{1}{\phi^2} = \frac{1}{\phi^3} + \frac{1}{\phi^4}$$

$$\frac{1}{\phi^3} = \frac{1}{\phi^4} + \frac{1}{\phi^5}$$

$$\cdots\cdots\cdots\cdots$$

$$\frac{1}{\phi^n} = \frac{1}{\phi^{n+1}} + \frac{1}{\phi^{n+2}} \quad , \qquad n = 0,1,2,3,\ldots\ldots$$

We will check now the first two equations above using MATLAB. The first equation $1 = \frac{1}{\phi} + \frac{1}{\phi^2}$ is checked using MATLAB as follows:

```
>> phi = (1+sqrt(5))/2

phi =

    1.6180

>> 1/phi + 1/phi^2

ans =

    1
```

Next, the second equation $\frac{1}{\phi} = \frac{1}{\phi^2} + \frac{1}{\phi^3}$ is checked as follows using MATLAB:

```
>> phi = (1+sqrt(5))/2

phi =
```

```
     1.6180

>> 1/phi^2 + 1/phi^3

ans =

     0.6180

>> 1/phi

ans =

     0.6180
```

Similarly, the other equations can be verified using MATLAB. There is another set of equations for the inverse powers of ϕ that is interesting. The following equations will be verified to be true:

$$\frac{1}{\phi^2} = 1 - \frac{1}{\phi}$$

$$\frac{1}{\phi^3} = \frac{2}{\phi} - 1$$

$$\frac{1}{\phi^4} = 2 - \frac{3}{\phi}$$

$$\frac{1}{\phi^5} = \frac{5}{\phi} - 3$$

$$\frac{1}{\phi^6} = 5 - \frac{8}{\phi}$$

$$\frac{1}{\phi^7} = \frac{13}{\phi} - 8$$

We can see that a pattern emerges for the expressions of the inverse powers of ϕ shown above. It is clear that the numbers appearing in the expressions on the right-hand-side of the equations are Fibonacci numbers. Thus, the above equations serve to strengthen the relationship between Fibonacci numbers and the Golden Ratio. Let us now check the equations for $\dfrac{1}{\phi^3}$ and $\dfrac{1}{\phi^4}$ shown above using MATLAB. First, the expression for $\dfrac{1}{\phi^3}$ is verified as follows:

```
>> phi = (1+sqrt(5))/2

phi =

    1.6180

>> 1/phi^3

ans =

    0.2361

>> 2/phi -1

ans =

    0.2361
```

Next, the expression for $\dfrac{1}{\phi^4}$ is verified using MATLAB as follows:

```
>> phi = (1+sqrt(5))/2
phi =
```

```
   1.6180

>> 1/phi^4

ans =

   0.1459

>> 2 - 3/phi

ans =

   0.1459
```

Similarly, the other equations shown above for the inverse powers of ϕ can be verified using MATLAB.

Next, we will derive the expressions for the previous two equations for $\dfrac{1}{\phi^3}$ and $\dfrac{1}{\phi^4}$ mathematically as follows. Let us start with the expression $\dfrac{1}{\phi} = \dfrac{1}{\phi^2} + \dfrac{1}{\phi^3}$ and substitute $1 - \dfrac{1}{\phi}$ for $\dfrac{1}{\phi^2}$ to obtain $\dfrac{1}{\phi^3} = \dfrac{1}{\phi} - \dfrac{1}{\phi^2} = \dfrac{1}{\phi} - \left(1 - \dfrac{1}{\phi}\right) = \dfrac{2}{\phi} - 1$. Similarly, we start with the expression $\dfrac{1}{\phi^2} = \dfrac{1}{\phi^3} + \dfrac{1}{\phi^4}$ and substitute $\dfrac{2}{\phi} - 1$ for $\dfrac{1}{\phi^3}$, and $1 - \dfrac{1}{\phi}$ for $\dfrac{1}{\phi^2}$ to obtain $\dfrac{1}{\phi^4} = \dfrac{1}{\phi^2} - \dfrac{1}{\phi^3} = \left(1 - \dfrac{1}{\phi}\right) - \left(\dfrac{2}{\phi} - 1\right) = 2 - \dfrac{3}{\phi}$.

We can continue in a similar fashion to prove the other expressions shown above and generate the Fibonacci numbers in this way.

There are more interesting properties of the Golden Ratio ϕ. For example, we can verify the following identities involving ϕ:

$$\frac{1}{\phi} + \frac{1}{\phi^2} = 1$$

$$\phi + \frac{1}{\phi^2} = 2$$

$$\phi^2 + \frac{1}{\phi^2} = 3$$

$$\left(\phi + \frac{1}{\phi^2}\right)^2 = 4$$

$$\left(\phi + \frac{1}{\phi}\right)^2 = 5$$

The above five identities are verified using MATLAB as follows:

```
>> phi = (1+sqrt(5))/2

phi =

    1.6180

>> 1/phi + 1/phi^2

ans =

     1
```

```
>> phi + 1/phi^2

ans =

      2

>> phi^2 + 1/phi^2

ans =

      3

>> (phi + 1/phi^2)^2

ans =

      4

>> (phi + 1/phi)^2

ans =

    5.0000
```

Another interesting property is that $\phi + \dfrac{1}{\phi} = \sqrt{5}$. This property is verified using MATLAB as follows:

```
>> phi = (1+sqrt(5))/2

phi =

    1.6180

>> phi + 1/phi
```

```
ans =

    2.2361

>> sqrt(5)

ans =

    2.2361
```

Finally, we list the following very interesting property of ϕ:

$$\phi = \frac{1}{\phi} + \frac{1}{\phi^2} + \frac{1}{\phi^3} + \frac{1}{\phi^4} + \frac{1}{\phi^5} + \frac{1}{\phi^6} + \ldots\ldots$$

We verify the above property using MATLAB as follows:

```
>> phi = (1+sqrt(5))/2

phi =

    1.6180

>> 1/phi + 1/phi^2 + 1/phi^3 + 1/phi^4 + 1/phi^5 + 1/phi^6

ans =

    1.5279
```

Obviously, we did not obtain the exact result above but we obtained an approximate value. This is because we used only the first six terms in the series. We would have obtained more precision and a more exact value for ϕ if we used more terms in the series.

The Golden Ratio can also be written as a continued fraction as follows

$$\phi = 1 + \cfrac{1}{1 + \cfrac{1}{1 + \cfrac{1}{1 + ...}}}$$

The above formula can be show to be valid easily as follows. It is very clear that the main denominator of the fraction on the right-hand-side is equal to ϕ also. Therefore, the above equation for the continued fraction is in the form $\phi = 1 + \dfrac{1}{\phi}$. This equation when simplified reduces to the familiar quadratic equation $\phi^2 = \phi + 1$ that is associated with the Golden Ratio.

We devote the remaining part of this chapter to deriving a compact equation for the computation of Fibonacci numbers using the value of the Golden Ratio. Let us start with the quadratic equation for the Golden Ratio that was derived in the previous chapter:

$$x^2 = x + 1$$

It was illustrated in the previous chapter that the two roots of the above equation are ϕ and $1 - \phi$. Substituting each value in the above quadratic equation separately, we obtain the following two equations:

$$\phi^2 = \phi + 1$$

$$(1 - \phi)^2 = (1 - \phi) + 1$$

Next, we multiply both sides of the first equation above by ϕ^n, and multiply both sides of the second equation above by $(1 - \phi)^n$ to obtain the following two equations:

$$\phi^{n+2} = \phi^{n+1} + \phi^n$$

$$(1-\phi)^{n+2} = (1-\phi)^{n+1} + (1-\phi)^n$$

where $n = 0,1,2,3,....$. Next, we subtract the second equation from the first equation above to obtain:

$$\phi^{n+2} - (1-\phi)^{n+2} = \phi^{n+1} - (1-\phi)^{n+1} + \phi^n - (1-\phi)^n$$

Next, we divide both sides of the above equation by $\phi - (1-\phi)$[23] to obtain the following recursive relation:

$$\frac{\phi^{n+2} - (1-\phi)^{n+2}}{\phi - (1-\phi)} = \frac{\phi^{n+1} - (1-\phi)^{n+1}}{\phi - (1-\phi)} + \frac{\phi^n - (1-\phi)^n}{\phi - (1-\phi)}$$

The above equation is a recursive relation which can be written in the general mathematical notation:

$$F(n+2) = F(n+1) + F(n)$$

It is seen that the above recursive relation is precisely the equation that governs the sequence of Fibonacci numbers (see the chapter on Fibonacci numbers). Therefore, by comparing the above two equations above, we can conclude that the general formula for $F(n)$ can be written as follows:

$$F(n) = \frac{\phi^n - (1-\phi)^n}{\phi - (1-\phi)} \quad , \quad n = 0,1,2,3,....$$

The above general equation can be then simplified as follows:

---

[23] This expression is equal to $2\phi - 1$ but will keep it in this unsimplified form in order to derive the sought relationship between Fibonacci numbers and the Golden Ratio.

$$F(n) = \frac{\phi^{n} - (1-\phi)^{n}}{2\phi - 1} \quad , \quad n = 0,1,2,3,\ldots$$

Thus, we conclude that the above equation can be used to computer Fibonacci numbers using the value of the Golden Ratio, i.e. the n-th term $F(n)$ of the Fibonacci sequence can be computed using the above general relation using the value of ϕ only. We can simplify the above general equation further by substituting the value $\dfrac{1+\sqrt{5}}{2}$ for ϕ in order to obtain:

$$F(n) = \frac{\left(1+\sqrt{5}\right)^{n} - \left(1-\sqrt{5}\right)^{n}}{2^{n}\sqrt{5}} \quad , \quad n = 0,1,2,3,\ldots.$$

Thus, the above formula is the general expression for computing any Fibonacci number without generating the whole sequence. In order to use the above equation, we only need to insert the value of n, i.e. we only need to know which Fibonacci number we need to compute.

Finally, we will write a short MATLAB function which uses the above formula to compute any Fibonacci number. The MATLAB code for the function `fib(n)` is shown below:

```
function f=fib(n)
% This function generates the nth Fibonacci
%  number
f      =      ((1+sqrt(5))^n     -(1-sqrt(5))^n)
/(sqrt(5)*2^n);
```

Next, let us execute the above MATLAB function several times in order to computer a few Fibonacci numbers. Here is the MATLAB output:

```
>> fib(5)

ans =

     5

>> fib(7)

ans =

     13

>> fib(10)

ans =

    55.0000

>> fib(15)

ans =

   610.0000
```

Thus, the MATLAB function `fib(n)` works perfectly. Just insert any value for n and execute the function `fib(n)` to get the desired nth term of the Fibonacci sequence without generating the sequence itself. We have been able to do this because we used the Golden Ratio.

It should be mentioned finally that there is an approximation for the above formula for calculating Fibonacci numbers.

Instead of using the above long formula, we can use the following shorter approximate formula:

$$F(n) \approx \frac{\phi^n}{\sqrt{5}}$$

Substituting the value of ϕ which is $\dfrac{1+\sqrt{5}}{2}$ in the above formula, we obtain the following explicit expression for the approximate formula:

$$F(n) \approx \frac{\left(1+\sqrt{5}\right)^n}{2^n\sqrt{5}}$$

It should be noted that the above approximate formula gives more accurate results for large value of n, i.e. for large Fibonacci numbers. The approximate short formula can be derived from the exact long formula by taking the limit[24] of the exact formula as n approaches infinity. We now write a new MATLAB function called approximate_fib(n) to calculate Fibonacci numbers using the approximate formula. Here is the MATLAB code for this function:

```
function f=approximate_fib(n)
% This function generates the nth Fibonacci
%  number using the approximate formula
f = ((1+sqrt(5))^n) /(sqrt(5)*2^n);
```

We execute the MATLAB function approximate_fib(n) for the first five n values 1, 2, 3, 4, 5 to obtain the following approximate values for Fibonacci numbers:

---

[24] Note that $\displaystyle\lim_{n\to\infty} \frac{\phi^n - \left(1-\phi\right)^n}{2\phi-1} = \frac{\phi^n}{\sqrt{5}}$. You need to revise your calculus information in order to derive this limit.

```
>> approximate_fib(1)

ans =

    0.7236

>> approximate_fib(2)

ans =

    1.1708

>> approximate_fib(3)

ans =

    1.8944

>> approximate_fib(4)

ans =

    3.0652

>> approximate_fib(5)

ans =

    4.9597
```

Next, we execute the above MATLAB function several times and compare its results with those of the exact function `fib(n)` as follows (we see that we obtain exact results for higher values of n):

```
>> fib(10)

ans =
```

```
    55.0000

>> approximate_fib(10)

ans =

    55.0036

>> fib(30)

ans =

  8.3204e+005

>> approximate_fib(30)

ans =
  8.3204e+005
```

Thus, for the tenth Fibonacci number, we obtain almost an exact result. We obtain also an almost exact result for the 30th Fibonacci number as shown above. The larger the value of n, the more accurate is the Fibonacci number which is obtained.

In the next chapter, we study another sequence of interesting numbers that is closely related to the Fibonacci sequence.

9. Lucas Numbers

In this chapter we will study Lucas numbers and the Lucas sequence of numbers. These numbers are closely associated with Fibonacci numbers. Lucas numbers are generated using the same rule as that of Fibonacci numbers, i.e. the next number is obtained by adding the previous two numbers. However, instead of starting with the numbers 1 and 1, we start with the numbers 2 and 1. Thus the third Lucas number is 3. The fourth Lucas number is 4. The fifth Lucas number is 7. Thus, the first few Lucas numbers are 2, 1, 3, 4, 7, 11, 18, 29, 47, 76, 123, …. This sequence of numbers is called the Lucas sequence. These numbers are generated using MATLAB interactively as follows:

```
>> 2+1

ans =

     3

>> 1+3

ans =

     4

>> 3+4

ans =

     7

>> 4+7

ans =
```

```
       11

>> 7+11

ans =

      18

>> 11+18

ans =

      29

>> 18+29

ans =

      47

>> 29+47

ans =

      76

>> 47+76

ans =

     123
```

In order to generate Lucas numbers automatically, here is the MATLAB code for the MATLAB function lucas(n) which generates the first n Lucas numbers.

```
function f=lucas(n)
```

```
% This function generates the first n Lucas
%  numbers
f = zeros(n,1);
f(1)=1;
f(2)=3;
for k = 3:n
    f(k) = f(k-1) + f(k-2);
end
```

Next, we execute the MATLAB function lucas(n) several times to generate a few Lucas numbers. The results are shown below:

```
>> lucas(7)

ans =

     1
     3
     4
     7
    11
    18
    29

>> lucas(20)

ans =

     1
     3
     4
     7
    11
    18
    29
    47
    76
```

```
        123
        199
        322
        521
        843
       1364
       2207
       3571
       5778
       9349
      15127
```

The above results show the first seven and twenty Lucas numbers, respectively. Using mathematical notation, let us denote the nth number in the Lucas sequence by L(n). Then, the Lucas sequence of numbers can be generated by using the following recursive relation:

$$L(n) = L(n-1) + L(n-2)$$

$$L(1) = 2$$

$$L(2) = 1$$

$$n = 3, 4, 5, 6, \ldots$$

The above relations for generating Lucas numbers are exactly the same as those of the Fibonacci numbers except the initial conditions, i.e. the two starting numbers, are different.

Let us plot a graph showing the relationship between the number of terms and the corresponding Lucas numbers. In this first plot, let us use only the first ten terms of the Lucas sequence. First, let us generate the x-axis of the plot as follows:

```
>> x = 1:10
```

```
x =

     1     2     3     4     5     6     7
8     9    10
```

The next step would be to generate the y-axis which would be the first ten Lucas numbers. These are generated and stored in the variable y as follows:

```
>> y = lucas(10)

y =

     1
     3
     4
     7
    11
    18
    29
    47
    76
   123
```

The final step would be to use the MATLAB command plot in order to generate the required graph. In addition, we use the MATLAB commands xlabel and ylabel to label the two axes, respectively. These commands are executed as shown below and the graph is show in Figure 9.1.

```
>> plot(x,y)
>> hold on
>> xlabel('n')
>> ylabel('lucas(n)')
```

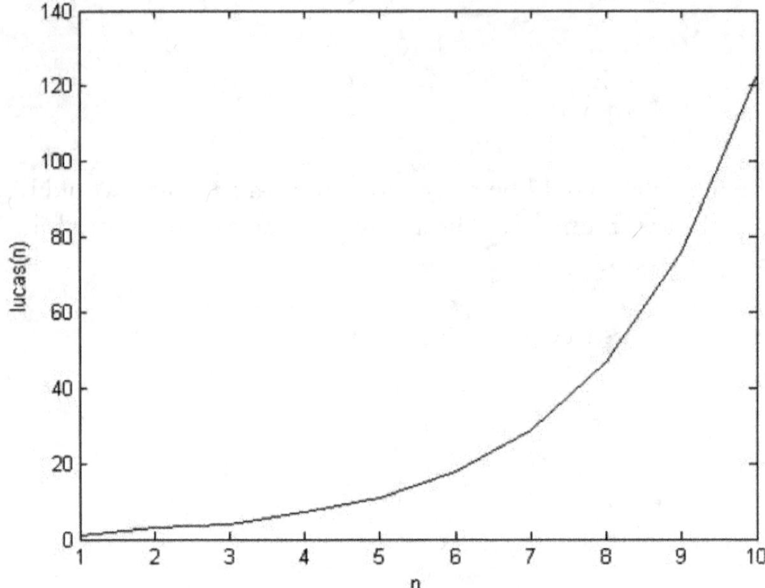

Figure 9.1: Graph of the first ten Lucas numbers

Similarly, the following commands will generate and plot the first 30 Lucas numbers. The plot is generated after the commands are executed sequentially and is shown in Figure 9.2. Each command is followed by semi-colon in order to suppress the listed output:

```
>> x = 1:30;
>> y = lucas(30);
>> plot(x,y)
>> hold on;
>> xlabel('n')
>> ylabel('lucas(n)')
```

It is seen from the graph that Lucas numbers increase exponentially as the value of n increases.

Using the mathematical notation of this chapter, the two graphs in Figures 9.1 and 9.2 are called $L(n)$ *vs. n graphs*. Next, we will study the graphs obtained when plotting the logarithms of Lucas numbers.

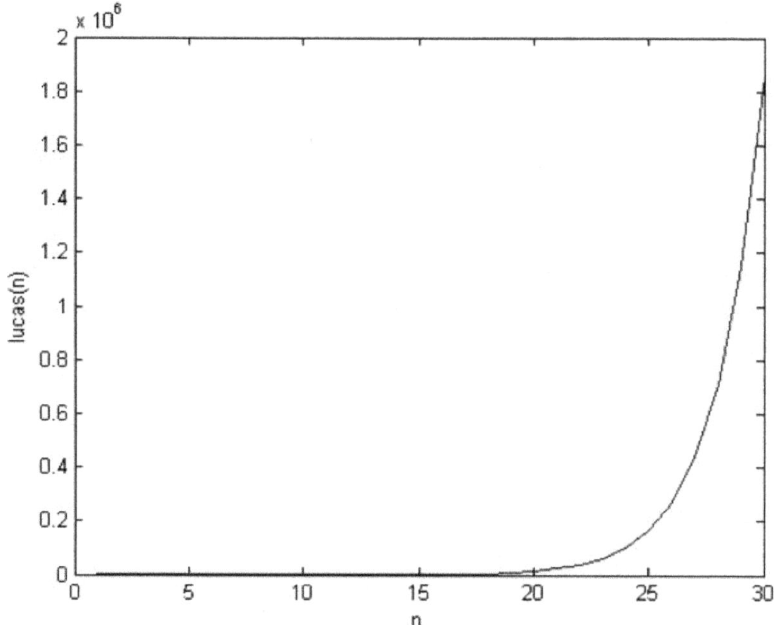

Figure 9.2: Graph of the first thirty Lucas umbers

First, let us use the MATLAB command `log` to calculate the logarithms[25] of the first ten Lucas numbers as follows:

```
>> log(lucas(10))

ans =

         0
    1.0986
    1.3863
    1.9459
    2.3979
    2.8904
    3.3673
```

---

[25] The logarithms calculated here are the natural logarithms, which are logarithms to the base e, where e = 2.71828... .

```
3.8501
4.3307
4.8122
```

We need now to plot the above logarithms on the y-axis while keeping the number of terms n on the x-axis. The type of plot obtained is called a semi-log graph. We will perform the following operations (while suppressing the output) in order to accomplish this. The resulting graph is shown in Figure 9.3. The graph is called a *log(L(n)) vs. n graph*. The MATLAB commands are shown below:

```
>> x = 1:10;
>> y = log(lucas(10));
>> plot(x,y)
>> hold on
>> xlabel('n')
>> ylabel('log(L(n))')
```

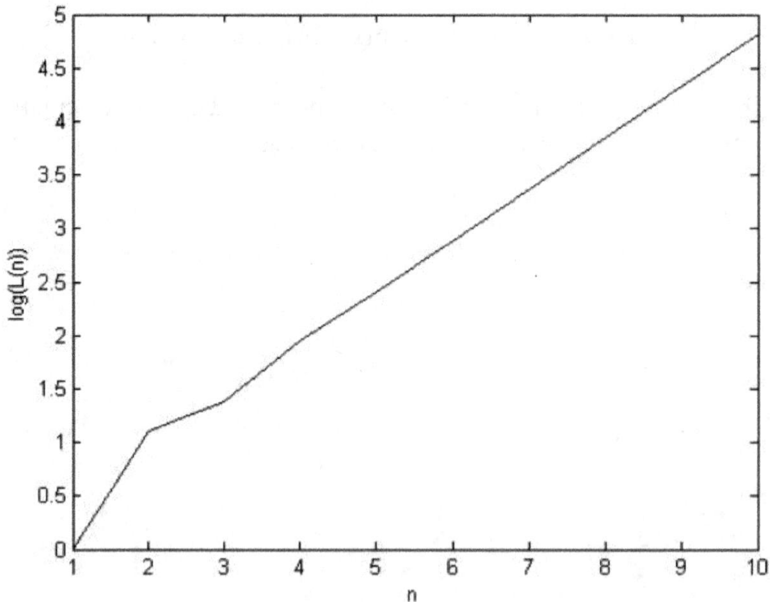

Figure 9.3: Semi-log graph of the first ten Lucas numbers

It is seen from Figure 9.3 that the graph is almost a straight line for values of n larger than 3. Let us calculate the slope[26] of this line. Let us perform this calculation by taking the fifth and tenth Lucas numbers and calculating the slope of the line segment between them. This is implemented in MATLAB as follows:

```
>> m = (y(10)-y(5))/(x(10)-x(5))

m =

    0.4829
```

Thus the slope of the straight line is 0.4829. In order to find the angle, we calculate the inverse tangent value of this number. This is obtained as $\tan^{-1}(0.4829) = 25.8°$.

Let us perform the above operations again but using the first 30 Lucas numbers. Here are the MATLAB commands to generate the plot while the graph is shown in Figure 9.4.

```
>> x = 1:30;
>> y = log(lucas(30));
>> plot(x,y)
>> hold on
>> xlabel('n')
>> ylabel('log(L(n))')
```

As expected, the semi-log graph shows almost a straight line. Let us now calculate the slope of this line between the 10th and the 30th Lucas numbers. This operation is implemented in MATLAB as follows

```
>> m = (y(30)-y(10))/(x(30)-x(10))
```

---

[26] The slope m of a line segment between two points (x_1,y_1) and (x_2,y_2) is calculated using the formula $m = (y_2-y_1)/(x_2-x_1)$.

```
m =

    0.4812
```

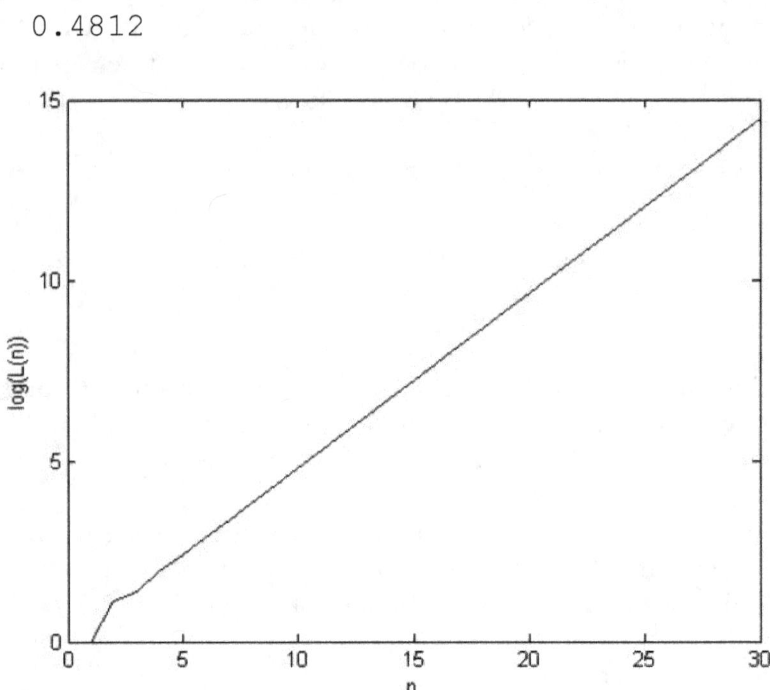

Figure 9.4: Semi-log graph of the first 30 Lucas numbers

As expected, the slope of the line is almost as before. The angle of the line is calculated as $\tan^{-1}(0.4812) = 25.7°$, almost exactly as before. It should also be noted that the slope and angle of the Lucas numbers are exactly the same as the slope and angle of Fibonacci numbers (see Chapter 6). It seems that this slope (or angle) is a constant that is associated with these types of sequences. An alternative method to plot the logarithms of Lucas numbers is to use the MATLAB command `semilogy` in association with the MATLAB function `lucas(n)`. The graph obtained in this way will also be an almost straight line but will be slightly different from the graphs obtained here.

Next, we will illustrate that the limiting ratio of the Lucas sequence of numbers is the Golden Ratio as in the case of the Fibonacci sequence. To illustrate this, let us calculate the ratio of each two consecutive numbers in the Lucas sequence.

Here are the first few terms of the Lucas sequence:

2, 1, 3, 4, 7, 11, 18, 29, 47, 76, 123, …

The ratio of each two consecutive Lucas numbers is calculated using MATLAB as follows:

```
>> 1/2

ans =

    0.5000

>> 3/1

ans =

    3

>> 4/3

ans =

    1.3333

>> 7/4

ans =

    1.7500

>> 11/7
```

```
ans =
    1.5714

>> 18/11

ans =
    1.6364

>> 29/18

ans =

    1.6111

>> 47/29

ans =

    1.6207

>> 76/47

ans =

    1.6170

>> 123/76

ans =

    1.6184

>> 199/123

ans =
```

```
    1.6179

>> 322/199

ans =

    1.6181
>> 521/322

ans =

    1.6180

>> 843/521

ans =

    1.6180
```

It is seen from the above MATLAB computations that the limiting ratio of the Lucas sequence approaches the number 1.6180 which is the Golden Ratio, exactly as in the case of the Fibonacci sequence. Next, a MATLAB function is shown which calculates the ratio as above and plots the graph of the ratio as the number of terms increases. Here is the code of the MATLAB function `lucas_ratio(n)` which calculates the limiting ratio:

```
function r=lucas_ratio(n)
% This function calculates the ratio of two
% consecutive Lucas numbers
f = zeros(n,1);
r = zeros(n,1);
f(1)=1;
f(2)=3;
for k = 3:n
    f(k) = f(k-1) + f(k-2);
    r(k) = f(k)/f(k-1);
```

```
end
```

Let us execute the above function for the first fifteen terms of the Lucas sequence. Here is output of the command:

```
>> lucas_ratio(15)

ans =

          0
          0
     1.3333
     1.7500
     1.5714
     1.6364
     1.6111
     1.6207
     1.6170
     1.6184
     1.6179
     1.6181
     1.6180
     1.6180
     1.6180
```

It is seen from the above output that the ratio of two consecutive Lucas numbers approaches the Golden Ratio which is the number 1.6180. To get more precision, we will use the `format long` command and execute the function again but with thirty terms as follows:

```
>> format long
>> lucas_ratio(30)

ans =

          0
```

```
                         0
1.33333333333333
1.75000000000000
1.57142857142857
1.63636363636364
1.61111111111111
1.62068965517241
1.61702127659574
1.61842105263158
1.61788617886179
1.61809045226131
1.61801242236025
1.61804222648752
1.61803084223013
1.61803519061584
1.61803352967830
1.61803416409969
1.61803392177224
1.61803401433308
1.61803397897799
1.61803399248243
1.61803398732419
1.61803398929446
1.61803398854189
1.61803398882935
1.61803398871955
1.61803398876149
1.61803398874547
1.61803398875159
```

It is seen from the above results that the limiting ratio approaches the value 1.6180339887... (accurate to ten decimal digits). Let us plot a graph of this ratio using the following commands where the output is suppressed:

```
>> x = 1:30;
>> y = lucas_ratio(30);
```

```
>> plot(x,y)
>> hold on
>> xlabel('n')
>> ylabel('L(n)/L(n-1)')
```

The resulting graph is shown in Figure 9.5. It is seen that the ratio L(n)/L(n-1) approaches the limiting value 1.61803.... which is the Golden Ratio, exactly as in the case of the Fibonacci sequence (see Chapter 7). The graph shown in the figure is know as the *L(n)/L(n-1) vs. n graph*.

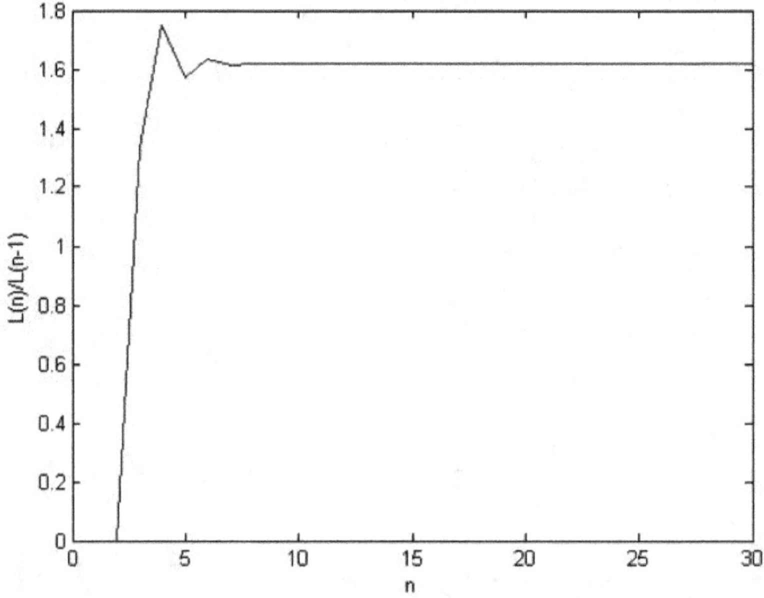

Figure 9.5: Graph of the limiting ratio of the Lucas sequence (shown for the first thirty terms of the sequence). The limiting ratio clearly approaches the value 1.61803...

Next, we will show several equations that will illustrate the relation between Fibonacci numbers and Lucas numbers. Using the terminology employed in this book, i.e. using F(n) to denote the nth

Fibonacci number and L(n) to denote the nth Lucas number, let us consider the following equation:

$$L(n+1) = F(n-1) + F(n+1) \quad , \quad n = 2, 3, 4, \ldots$$

To check the validity of the above equation, let us substitute several values for n. For n = 2, we obtain L(3) = F(1) + F(3). Looking at the Fibonacci and Lucas sequences below, it can be seen that the above equation reduces to 3 = 1 + 2 which is correct.

Fibonacci sequence: 1, 1, 2. 3, 5, 8, 13, 21, …...
Lucas sequence: 2, 1, 3, 4, 7, 11, 18, 29, ….

Let us check the above equation for n = 3. In this case, we obtain L(4) = F(2) + F(4). Using the Fibonacci and Lucas sequences above, we obtain 4 = 1 + 3 which is correct. Let us now check the above equation for n = 5. Using the Fibonacci and Lucas sequences above, we obtain L(6) = F(4) + F(6) which reduces to 11 = 3 + 8 which is correct. This proves[27] the validity of the above relation between Fibonacci and Lucas numbers.

There are still more relations between the two sequences of Fibonacci and Lucas. For example, consider the following relation:

$$L^2(n+1) = 5F^2(n) + 4(-1)^n \quad , \quad n = 1, 2, 3, \ldots$$

Let us now check the validity of the above equation by substituting several values for n along with the Fibonacci and Lucas numbers shown above. For example, for n = 1, we obtain $L^2(2) = 5F^2(1) + 4(-1)^1$. This reduces to $1^2 = 5(1)^2 + (-4)$. This finally becomes $1 = 5 - 4$ which is correct.

Let us now check the validity of the above equation for n = 4. For this value, the above equation becomes $L^2(5) = 5F^2(4) + 4(-1)^4$.

---

[27] This does not constitute a formal proof; we are just showing that the equation is correct for some values of n. For a formal proof, we should use the method of mathematical induction but this is beyond the scope of this work.

When substituting the relevant Fibonacci and Lucas numbers, the equation reduces to $7^2 = 5(3)^2 + 4(-1)^4$. This finally becomes $49 = 45 + 4$ which is correct. This proves the validity of the above relation between the two sequences.

Another relation between Fibonacci and Lucas numbers is the following equation:

$$F(2n) = L(n+1)\, F(n) \quad , \quad n = 1, 2, 3, \ldots$$

Let us check the validity of the above equation. For n = 1, the equation becomes $F(2) = L(2)F(1)$ which reduces to $1 = (1)(1)$ which is correct. For n = 3, we obtain $F(6) = L(4)F(3)$ which reduces to $8 = (4)(2)$ which is correct.

The final relation we will consider between Fibonacci and Lucas numbers is the following equation:

$$5F(n) = L(n) + L(n+2) \quad , \quad n = 1, 2, 3, \ldots\ldots$$

Let us check the validity of the above equation. For n = 1, the equation becomes $5F(1) = L(1) + L(3)$ which reduces to $5 = 2 + 3$ which is correct. For n = 4, we obtain $5F(4) = L(4) + L(6)$ which reduced to $5(3) = 4 + 11$. This is correct. This proves the validity of the above relation between the two sequences.

In the above, we have illustrated the relationship between the Fibonacci and Lucas sequences through four distinct equations. Other equations can be written relating the two sequences but the four equations given above will suffice.

Let us now show how to compute the value of the nth Lucas number directly without listing the preceding Lucas numbers in the sequence. We will show this in a way analogous to that used for the Fibonacci sequence in Chapter 8. The nth Lucas number in the

Lucas sequence is given directly by the following formula[28] in terms of the Golden Ratio ϕ:

$$L(n) = \phi^{n-1} + (1-\phi)^{n-1} \quad , \quad n = 1,2,3,.....$$

For example, let us calculate the eighth Lucas number using the above equation. Here are the MATLAB commands for this calculation using the formula given above:

```
>> phi = (1+sqrt(5))/2

phi =

    1.6180

>> L8 = phi^7 + (1-phi)^7

L8 =

    29.0000
```

Therefore, we get the value of 29 as the eighth Lucas number which is correct. Here is the code of the MATLAB function `luc(n)` that implements the above formula for the calculation of the nth Lucas number:

```
function f=luc(n)
% This function generates the nth Lucas number
f = ((1+sqrt(5))/2)^(n-1) + (1-(1+sqrt(5))/2)^(n-
1);
```

Let us execute the above function several times to compute several distinct Lucas numbers directly. The results are shown below:

```
>> luc(3)
```

---

[28] The derivation of this formula is left to the reader. The derivation can easily be made in a way similar to the derivation of the formula for the nth Fibonacci number that was detailed in Chapter 8.

```
ans =

     3

>> luc(8)

ans =

    29.0000

>> luc(15)
ans =

   843.0000
```

The advantage of using the MATLAB function `luc(n)` over using the MATLAB function `lucas(n)` is that there is now no need to generate the entire Lucas sequence in order to calculate the desired Lucas number. In a way similar to the formula given in Chapter 8 for Fibonacci numbers, there is also an approximate formula for the nth Lucas number given as follows:

$$L(n) \approx \phi^{n-1}$$

The above approximate formula is implemented in the MATLAB function `approximate_luc(n)` whose code is shown below:

```
function f=approximate_luc(n)
% This function generates the nth Lucas number
% using the approximate formula
f = ((1+sqrt(5))/2)^(n-1);
```

Let us execute the above function several times to find approximate values for several Lucas numbers. The results are shown below:

```
>> approximate_luc(3)

ans =

    2.6180

>> approximate_luc(8)

ans =

    29.0344
>> approximate_luc(15)

ans =

   842.9988
```

The above results for the approximate values of Lucas numbers can be compared with the exact results for the same Lucas numbers given before.

In the next chapter, we will investigate other types of sequences that are similar to the Fibonacci and Lucas sequences.

MATLAB Guide to Fibonacci Numbers and the Golden Ratio
A Simplified Approach, for Students and Scientists and Engineers

10. Generalizations of Fibonacci Numbers

In this chapter we will study several methods to generalize Fibonacci numbers. One way is to change the initial conditions, i.e. change the first two numbers in the sequence. This method was illustrated in Chapter 9 when we studied the Lucas sequence. The Lucas sequence was obtained from the Fibonacci rule by changing the first two terms of the Fibonacci sequence. Another method to generalize Fibonacci numbers is to use the sum of the preceding three terms instead of the preceding two terms in order to generate the next term in the sequence. Both these methods will be illustrated in this chapter.

A Fibonacci sequence of order n is a sequence of numbers in which each number is the sum of the previous n numbers. Thus, the usual Fibonacci numbers form a Fibonacci sequence of order 2. This is because each number of the Fibonacci sequence is the sum of the previous two numbers.

Extending the above rule to the number 3, we can generate the tribonacci sequence of numbers. In the tribonacci sequence, each number is the sum of the previous three numbers. Thus, the tribonacci sequence starts with the numbers 0, 0, 1, 1, 2, 4, 7, 13, 24, 44, 81, 149, ... The following is the MATLAB code for the function `tribonacci(n)` which generates the first n tribonacci numbers:

```
function f=tribonacci(n)
% This function generates the first n
tribonacci numbers
   f = zeros(n,1);
   f(1)=0;
   f(2)=0;
   f(3)=1;
   for k = 4:n
```

```
      f(k) = f(k-1) + f(k-2) + f(k-3);
end
```

Let us run the above function to generate the first twenty tribonacci numbers as follows:

```
>> tribonacci(20)

ans =

            0
            0
            1
            1
            2
            4
            7
           13
           24
           44
           81
          149
          274
          504
          927
         1705
         3136
         5768
        10609
        19513
```

In the tribonacci sequence, the ratio toward which adjacent tribonacci numbers tend is a constant similar to the Golden Ratio. This constant may be obtained by solving the following algebraic equation:

$$x^3 - x^2 - x - 1 = 0$$

The solution to the above equation is obtained as the following expression:

$$\frac{1+\sqrt[3]{19+3\sqrt{33}}+\sqrt[3]{19-3\sqrt{33}}}{3}$$

Evaluating the above expression, it is found that its numerical value is approximately $1.83929\ldots$.

Using MATLAB, the above algebraic equation is solved as follows:

```
>> solve('x^3 -x^2 -x -1=0');

>> double(ans)

ans =

    1.8393
   -0.4196 + 0.6063i
   -0.4196 - 0.6063i
```

The above constant may also be obtained by solving the following algebraic equation:

$$x + x^{-3} = 2$$

The tribonacci numbers may also be evaluated directly by using the following formula:

$$T(n) = 3b\,\frac{\dfrac{1}{3}\left(a_{+}+a_{-}+a\right)^{n}}{b^{2}-2b+4}$$

where the coefficients a_{+}, a_{-}, and b are given by:

$$a_{\pm} = \left(19 \pm 3\sqrt{33}\right)^{1/3}$$

$$b = \left(586 + 102\sqrt{33}\right)^{1/3}$$

Generalizing the above discussion to the number 4, we can generate the tetranacci sequence of numbers. Each tetranacci number is obtained as the sum of the previous four numbers in the tetranacci sequence. Thus, the first few terms in this sequence are 0, 0, 0, 1, 1, 2, 4, 8, 15, 29, 56, 108, 208, ……. .

The following MATLAB code is for the function tetranacci(n) which generates the first n tetranacci numbers:

```
function f=tetranacci(n)
% This function generates the first n
tetranacci numbers
f = zeros(n,1);
f(1)=0;
f(2)=0;
f(3)=0;
f(4)=1;
for k = 5:n
    f(k) = f(k-1) + f(k-2) + f(k-3) + f(k-
4);
end
```

We next run the above function as follows to generate the first twenty tetranacci numbers:

```
>> tetranacci(20)

ans =
                 0
                 0
                 0
                 1
                 1
                 2
                 4
                 8
```

```
      15
      29
      56
     108
     208
     401
     773
    1490
    2872
    5536
   10671
   20569
```

In this case, the ratio to which adjacent tetranacci numbers tend to is a constant and is obtained by solving the following algebraic equation:

$$x^4 - x^3 - x^2 - x - 1 = 0$$

The solution to the above equation is obtained approximately as 1.92756.... . This is called the tetranacci constant. The above equation is solved as follows using MATLAB:

```
>> solve('x^4 - x^3 -x^2 -x -1=0');

>> double(ans)
ans =

    1.9276
   -0.7748
   -0.0764 + 0.8147i
   -0.0764 - 0.8147i
```

By making similar generalizations to the numbers 5, 6, and 7, we can also obtain pentanacci numbers, hexanacci numbers, and heptanacci numbers, respectively.

Each pentanacci number is obtained as the sum of the previous five numbers in the pentanacci sequence. Thus, the first few terms of the pentanacci sequence are 0, 0, 0, 0, 1, 1, 2, 4, 8, 16, 31, 61, 120, ... Here is the MATLAB code for the function pentanacci(n) followed by a run of the function to generate the first twenty pentanacci numbers:

```
function f=pentanacci(n)
% This function generates the first n
pentanacci numbers
f = zeros(n,1);
f(1)=0;
f(2)=0;
f(3)=0;
f(4)=0;
f(5)=1;
for k = 6:n
    f(k) = f(k-1) + f(k-2) + f(k-3) + f(k-
4) +f(k-5);
end
```

```
>> pentanacci(20)

ans =

         0
         0
         0
         0
         1
         1
         2
         4
         8
        16
        31
        61
       120
       236
       464
       912
      1793
```

```
3525
6930
13624
```

Each hexanacci number is obtained as the sum of the previous six numbers in the hexanacci sequence. Thus, the first few terms of the hexanacci sequence are 0, 0, 0, 0, 0, 1, 1, 2, 4, 8, 16, 32, 63, 125, ... The following is the MATLAB code for the function hexanacci(n) followed by a run of the function to generate the first twenty hexanacci numbers:

```
function f=hexanacci(n)
% This function generates the first n
hexanacci numbers
f = zeros(n,1);
f(1)=0;
f(2)=0;
f(3)=0;
f(4)=0;
f(5)=0;
f(6)=1;
for k = 7:n
    f(k) = f(k-1) + f(k-2) + f(k-3) + f(k-
4) + f(k-5) +f(k-6);
end

>> hexanacci(20)

ans =

                    0
                    0
                    0
                    0
                    0
                    1
                    1
                    2
                    4
                    8
                   16
```

```
        32
        63
       125
       248
       492
       976
      1936
      3840
      7617
```

Each heptanacci number is obtained as the sum of the previous seven numbers in the heptanacci sequence. Thus, the first few terms of the heptanacci sequence are 0, 0, 0, 0, 0, 0, 1, 1, 2, 4, 8, 16, 32, 64, 127, ... The following is the MATLAB code for the function heptanacci(n) followed by a run of the function to generate the first twenty heptanacci numbers:

```
function f=heptanacci(n)
%  This  function  generates  the  first  n
heptanacci numbers
  f = zeros(n,1);
  f(1)=0;
  f(2)=0;
  f(3)=0;
  f(4)=0;
  f(5)=0;
  f(6)=0;
  f(7)=1;
  for k = 8:n
      f(k) = f(k-1) + f(k-2) + f(k-3) + f(k-
4) + f(k-5) +f(k-6) +f(k-7);
  end


>> heptanacci(20)

ans =

                0
                0
                0
```

$$0$$
$$0$$
$$0$$
$$1$$
$$1$$
$$2$$
$$4$$
$$8$$
$$16$$
$$32$$
$$64$$
$$127$$
$$253$$
$$504$$
$$1004$$
$$2000$$
$$3984$$

In general, a polynacci sequence would, after an infinite number of zeros, yield the following sequence of numbers:

$$[\ldots\ldots, 0, 0, 0, 1,]\ \ 1, 2, 4, 8, 16, 32, 64, 128, \ldots\ldots$$

The above sequence is simply the powers of 2.

In general, the k-th number of the n-nacci sequence of numbers is given directly by the following formula:

$$F_k^{(n)} = \frac{r^{k-1}(r-1)}{(n+1)r - 2n}$$

where r is the n-nacci constant which may be obtained by solving the following general algebraic equation:

$$x + x^{-n} = 2$$

In general, the following linear recurrence relation

$$F(n) = \sum_{i=1}^{k} F(n-i)$$

has a limiting ratio that may be obtained as the solution of the following general algebraic equation:

$$x^k - x^{k-1} - x^{k-2} - \ldots\ldots\ldots - x - 1 = 0$$

Finally, we use the above general linear recurrence relation and the associated algebraic equation to derive the following examples of generalized Fibonacci sequences:

Next, we show further examples. For more details, see the books in references [48-56] and the web links [57-67].

The sequence $F(n) = F(n-1) + F(n-2)$ has a limiting ratio of 1.618033… which is the Golden Ratio which is obtained as the solution to the algebraic equation $x^2 - x - 1 = 0$. Here is the MATLAB code to obtain the solution:

```
>> solve('x^2 - x - 1 = 0');

>> double(ans)

ans =

   1.6180
  -0.6180
```

The sequence $F(n) = F(n-1) + F(n-3)$ has a limiting ratio which is obtained as the solution to the algebraic equation $x^3 - x - 1 = 0$. Here is the MATLAB code to obtain the solution:

```
>> solve('x^3- x - 1 = 0');

>> double(ans)
```

ans =

```
   1.3247
  -0.6624 + 0.5623i
  -0.6624 - 0.5623i
```

The sequence $F(n) = F(n-2) + F(n-3)$ has a limiting ratio which is obtained as the solution to the algebraic equation $x^3 - x^2 - 1 = 0$. Here is the MATLAB code to obtain the solution:

```
>> solve('x^3- x^2 - 1 = 0');

>> double(ans)
```

ans =

```
   1.4656
  -0.2328 + 0.7926i
  -0.2328 - 0.7926i
```

The sequence $F(n) = F(n-1) + F(n-4)$ has a limiting ratio which is obtained as the solution to the algebraic equation $x^4 - x - 1 = 0$. Here is the MATLAB code to obtain the solution:

```
>> solve('x^4- x - 1 = 0');

>> double(ans)
```

ans =

```
   1.2207
  -0.7245
  -0.2481 + 1.0340i
  -0.2481 - 1.0340i
```

The sequence $F(n) = F(n-3) + F(n-4)$ has a limiting ratio which is obtained as the solution to the algebraic equation $x^4 - x^3 - 1 = 0$. Here is the MATLAB code to obtain the solution:

```
>> solve('x^4- x^3 - 1 = 0');

>> double(ans)

ans =

    1.3803
   -0.8192
    0.2194 + 0.9145i
    0.2194 - 0.9145i
```

The sequence $F(n) = F(n-1) + F(n-5)$ has a limiting ratio which is obtained as the solution to the algebraic equation $x^5 - x - 1 = 0$. Here is the MATLAB code to obtain the solution:

```
>> solve('x^5- x - 1 = 0');

>> double(ans)

ans =

    1.1673
    0.1812 + 1.0840i
   -0.7649 + 0.3525i
   -0.7649 - 0.3525i
    0.1812 - 1.0840i
```

The sequence $F(n) = F(n-2) + F(n-5)$ has a limiting ratio which is obtained as the solution to the algebraic equation $x^5 - x^2 - 1 = 0$. Here is the MATLAB code to obtain the solution:

```
>> solve('x^5- x^2 - 1 = 0');

>> double(ans)

ans =

   1.1939
   0.1546 + 0.8281i
  -0.7515 + 0.7846i
  -0.7515 - 0.7846i
   0.1546 - 0.8281i
```

The sequence $F(n) = F(n-3) + F(n-5)$ has a limiting ratio which is obtained as the solution to the algebraic equation $x^5 - x^3 - 1 = 0$. Here is the MATLAB code to obtain the solution:

```
>> solve('x^5- x^3 - 1 = 0');

>> double(ans)

ans =

   1.2365
   0.3408 + 0.7854i
  -0.9590 + 0.4284i
  -0.9590 - 0.4284i
   0.3408 - 0.7854i
```

The sequence $F(n) = F(n-4) + F(n-5)$ has a limiting ratio which is obtained as the solution to the algebraic equation $x^5 - x^4 - 1 = 0$. Here is the MATLAB code to obtain the solution:

```
>> solve('x^5- x^4 - 1 = 0');

>> double(ans)
```

ans =

```
  0.5000 + 0.8660i
  0.5000 - 0.8660i
  1.3247
 -0.6624 + 0.5623i
 -0.6624 - 0.5623i
```

For more details, see the books in references [48-56] and the web links [57-67]. In the next chapter, we investigate random Fibonacci sequences.

11. Random Fibonacci Numbers

In this chapter we will study random Fibonacci sequences briefly. In this case, we will obtain a different Fibonacci sequence each time because the numbers in the sequence are chosen according to a random rule. In this way, we will obtain random Fibonacci numbers. The general rule to generate such random sequences is the following recurrence relation:

$$F(n) = F(n-1) \pm F(n-2)$$

where the plus or minus sign is chosen at random with equal probability $\frac{1}{2}$, independently of the value of n. It has been demonstrated in the literature that this sequence generally grows exponentially.

In 1999, Divakar Viswanath showed that the growth rate of the random Fibonacci sequence is equal to 1.13198924.... . This number is actually the limiting ratio of any random Fibonacci sequence. This is a new mathematical constant and is now known as Viswanath's constant.

The random Fibonacci sequence starts with the two numbers 1 and 1, and the value of each subsequent number in the sequence is determined by a fair coin toss. Given two consecutive numbers of the sequence, the next number is either the sum of the two numbers or their difference with probability $\frac{1}{2}$, independently of the choice made previously.

Next, we show one example how to generate random Fibonacci sequences using MATLAB. Below is the MATLAB code for the function `random_fibonacci(n)` which generates n random Fibonacci numbers (note that the implementation of this function may be slightly different than the rule presented in this chapter).

```
function f=random_fibonacci(n)
% This function generates the first n
random Fibonacci numbers
f = zeros(n,1);
f(1)=1;
f(2)=2;
o = [-1,1];
s = o(ceil(2.0*rand(1,1)));
for k = 3:n
    f(k) = f(k-1) + s*f(k-2);
end
```

Next, we run the above function as follows in order to generate twenty random Fibonacci numbers:

```
>> random_fibonacci(20)

ans =

        1
        2
        1
       -1
       -2
       -1
        1
        2
        1
       -1
       -2
       -1
        1
        2
        1
       -1
       -2
       -1
```

1
2

Other variations of the above definition of random Fibonacci sequences also exist but are beyond the scope of this book.

We will not show how Viswanath's constant is obtained mathematically. This is beyond the scope of this work. The mathematics to generate this constant is somewhat involved and the reader is directed to Viswanath's paper for further details. See the books in references [48-56] and the web links [57-67].

MATLAB Guide to Fibonacci Numbers and the Golden Ratio
A Simplified Approach, for Students and Scientists and Engineers

References

MATLAB Tutorials

1. http://www.cyclismo.org/tutorial/matlab/

   This is an online tutorial in HTML format from Union College, New York. The tutorial emphasizes vectors, matrices, vector operations, loops, plots, executable files (scripts), subroutines (functions), if statements, and data files.

2. http://www.math.ufl.edu/help/matlab-tutorial/

   This is an online tutorial in HTML format from the University of Florida, Gainesville. The tutorial emphasizes matrices, variables, functions, decisions, loops, and scripts. This tutorial has a nice list of important commands of MATLAB with a short description of each command.

3. http://www.mines.utah.edu/gg_computer_seminar/matlab/matlab.html

   This is an online tutorial in HTML format available at the website of University of Utah, Salt Lake City (Original by Kermit Sigmon, University of Florida). This tutorial emphasizes matrices, decisions, loops, scalar functions, vector functions, matrix functions, strings, and graphics. You may also download the tutorial as a postscript file (39 pages that include a comprehensive reference section).

4. http://web.mit.edu/afs/.athena/astaff/project/logos/olh/Math/Matlab/Matlab.html

   This is an online tutorial in HTML format hosted at MIT. It emphasizes matrices, arithmetic and logical operators, control structures, selective indexing, polynomial operations, signal processing functions, graphics, scripts, and functions.

5.   http://www.mit.edu/people/abbe/matlab/main.html

This is an online 3-day tutorial at MIT in HTML format. It covers matrices, various MATLAB commands, help, plotting, polynomials, variables, scripts, functions, loops, debugging, differential equations, vectorization, three-dimensional graphics, and symbolic math.

6.   http://texas.math.ttu.edu/~gilliam/ttu/m4330/m4330.html

This is a tutorial hosted at Texas Tech University, Lubbock. They provide six PDF files for download in the form of six lessons for a mathematics course they teach there. These lessons are somewhat advanced and specialized and are not recommended for beginner students of MATLAB.

7.   http://www.engin.umich.edu/group/ctm/basic/basic.html

This is a basic tutorial from Carnegie Mellon and the University of Michigan. It covers the basics of vectors, functions, plotting, polynomials, matrices, printing, help, and M-files.

8.   http://www.mathworks.com/academia/student_center/tutorials/launchpad.html

This is an online tutorial from the MathWorks, the company that develops and sells MATLAB. This tutorial covers variables, calculations, plotting, scripts, and files.

9.   http://www.math.utah.edu/lab/ms/matlab/matlab.html#starting

This is an online short tutorial from the University of Utah, Salt Lake City. This is a very simple tutorial recommended for beginners. It covers matrices, vectors, systems of equations, loops, and graphing in one and two dimensions.

10. http://www.math.mtu.edu/~msgocken/intro/intro.html

This is an online tutorial in HTML format hosted at Michigan Tech and written by Mark S. Gockenback. It covers calculations, graphs, programming, advanced matrix calculations, advanced graphics, solving nonlinear problems, and advanced data types like structures, cell arrays, and objects.

11. http://www.eece.maine.edu/mm/matweb.html

MATLAB Educational Websites: This is an online HTML page containing links to several educational websites on various topics using MATLAB. The page contains also links to numerous MATLAB tutorials.

12. http://web.cecs.pdx.edu/~gerry/MATLAB/

MATLAB Hypertext Reference: This is an online tutorial in HTML format written by Gerald Recktenwald and hosted at Portland State University, Portland. This tutorial covers variables, plotting, and programming.

13. http://www.facstaff.bucknell.edu/maneval/help211/helpmain.html

Helpful Information for Using MATLAB: These pages of helpful information are maintained by Jim Maneval and hosted at Bucknell University, Lewisburg.

14. http://www.indiana.edu/~statmath/math/matlab/gettingstarted/index.html

Getting Started with MATLAB: This is an online tutorial in HTML format. The tutorial may also be downloaded and printed as a PDF file (14 pages). This tutorial covers syntax, matrices, graphics, and programming.

15. www.math.udel.edu/~driscoll/teaching/**matlab_adv.pdf**

Crash Course in MATLAB: This tutorial is in the form of a PDF file download. It is written by Tobin A. Driscoll and hosted at the University of Delaware, The tutorial is 66 pages that can be downloaded and printed. This tutorial covers arrays, matrices, scripts, functions, errors, 2D and 3D graphics, color, handles and properties, vectorization, advanced data structures (strings, cell arrays, structures), linear algebra, optimization, data fitting, quadrature, and differential equations.

16. www.maths.dundee.ac.uk/~ftp/na-reports/**MatlabNotes**.pdf

An Introduction to MATLAB: This tutorial is in the form of a PDF file download. It is written by David F. Griffiths and is hosted at the University of Dundee. The tutorial is 37 pages that can be downloaded and printed. This tutorial covers numbers, formats, variables, output, vectors, 2D plotting, scripts, vector operations, matrices, matrix operations, systems of linear equations, strings, loops, decisions, 3D plotting, files, and graphical user interfaces.

17. www.geosci.uchicago.edu/~gidon/geosci236/organize/**matlabIntro.pdf**

An Introduction to MATLAB: This tutorial is in the form of a PDF file download. It is written by S. Butenko, P. Pardalos, and L. Pitsoulis. It is hosted at the University of Chicago. The tutorial is 28 pages that can be downloaded and printed. This tutorial covers matrices, matrix operations, workspace, formats, functions, vector functions, matrix functions, polynomial functions, programming, loops, decisions, M-files, and 2D and 3D graphics.

18. http://www.math.colostate.edu/~gerhard/classes/340/notes/index.html

MATLAB Notes: This tutorial is in the form HTML pages online. It is hosted at Colorado State University. This tutorial covers arithmetic, array operations, scripts, functions, plotting, polynomials, matrices, vectors, linear systems of algebraic equations, and symbolic computation. There are some links to more tutorials on the internet that are hosted at other places.

19. http://www.indiana.edu/~statmath/math/matlab/gettingstarted/index.html

Getting Started with MATLAB: This tutorial is in the form of online HTML pages that are hosted at Indiana University. It may also be downloaded as a PDF file (14 pages) and printed. This tutorial covers syntax, matrices, graphics, and programming.

20. http://www-ccs.ucsd.edu/matlab/toolbox/symbolic/symbmath.html

This is an online reference guide for the MATLAB Symbolic Math Toolbox. It is hosted at the University of California, San Diego. The tutorial covers about 85 MATLAB symbolic math commands that are listed alphabetically with examples.

21. http://www.phys.ufl.edu/docs/matlab/toolbox/symbolic/ch1.html

Using the Symbolic Math Toolbox: This tutorial is in the form of HTML pages that is hosted at the University of

Florida, Gainesville. This tutorial seems to be comprehensive and covers calculus, simplifications, substitutions, variable-precision arithmetic, linear algebra, solving equations, special mathematical functions, using Maple functions, and the extended symbolic math toolbox.

22. http://en.wikipedia.org/wiki/Matlab

This is a small overview in the form of an HTML page hosted at the website of Wikipedia, the free internet encyclopedia. The page has a brief history of MATLAB and some useful links.

23. http://www.facstaff.bucknell.edu/maneval/help211/exercises.html

MATLAB Exercises: This web page provides some MATLAB exercises for practice. The site is hosted at Bucknell University. The exercises cover syntax, arrays, relational operators, logical operators, decisions, loops, and programming.

24. http://www.math.chalmers.se/~nilss/enm/INTRO_matlab_I.html

MATLAB Exercises: This web page provides three sets of MATLAB exercises for practice. The site is hosted at Chalmers University of Technology and Gteborg University. The exercises cover arithmetic, formats, variables, vectors, matrices, decisions, scripts, loops, computer arithmetic, strings, and graphics.

25. http://college.hmco.com/mathematics/larson/elementary_linear/5e/students/matlabs.html

MATLAB Exercises: These are MATLAB exercises taken from the book "Elementary Linear Algebra", fifth edition, by

Ron Larson. Each set of exercises is downloaded as a PDF file that can be printed. These exercises cover systems of linear equations, matrices, determinants, vector spaces, inner product spaces, linear transformations, eigenvalues and eigenvectors, and complex vector spaces.

26. http://www.math.umn.edu/~ROberts/math5385/matlabEx1.html

Supplementary MATLAB Exercises: These exercises are hosted at the University of Minnesota. They are written specifically for a mathematics course at the university. These exercises mostly deal with the graphics capabilities of MATLAB.

27. www.kom.aau.dk/~borre/**matlab**7/exercise.pdf

This is a PDF file download of a set of 8 pages of MATLAB exercises. The exercises are somewhat advanced and are not recommended for beginning students of MATLAB.

28. www.chaos.swarthmore.edu/courses/Phys50L_2006/**Matlab/MatlabExercises**.pdf

This is a PDF file download of 27 pages of MATLAB exercises. These exercises are written for a physics course and are somewhat advanced. They are not recommended for beginners.

MATLAB Books:

29. Gilat, A., *MATLAB: An Introduction with Applications*, Second Edition, John Wiley & Sons, 2004.

30. Pratap, R., *Getting Started with MATLAB 7: An Introduction for Scientists and Engineers*, Oxford University Press, 2005.

31. Hanselmann, D. and Littlefield, B., *Mastering MATLAB 7*, Prentice Hall, 2004.

32. Palm, W., *Introduction to MATLAB 7 for Engineers*, McGraw-Hill, 2004.

33. Moore, H., *MATLAB for Engineers*, Prentice Hall, 2006.

34. Chapman, S., *MATLAB Programming for Engineers*, Thomson Engineering, 2004.

35. Davis, T. and Sigmon, K., *MATLAB Primer*, Seventh Edition, Chapman & Hall, 2004.

36. Higham, D. and Higham, N., *MATLAB Guide*, Second Edition, SIAM, 2005.

37. King, J., *MATLAB for Engineers*, Addison-Wesley, 1988.

38. Etter, D., *Introduction to MATLAB for Engineers and Scientists*, Prentice Hall, 1995.

39. Magrab, E. et al., *An Engineer's Guide to MATLAB*, Prentice Hall, 2000.

40. Etter, D., Kuncicky, D. and Hull, D., *Introduction to MATLAB 6*, Prentice Hall, 2001.

41. Recktenwald, G., *Introduction to Numerical Methods and MATLAB: Implementation and Applications*, Prentice Hall, 2001.

42. Biran, A. and Breiner, M., *MATLAB 5 for Engineers*, Addison-Wesley, 1999.

43. Part-Enander, E. and Sjoberg, A., *The MATLAB 5 Handbook*, Addison-Wesley, 1999.

44. Etter, D., *Engineering Problem Solving with MATLAB*, Prentice Hall, 1993.

45. Chen, K., Giblin, P, and Irving, A., *Mathematical Explorations with MATLAB*, Cambridge University Press, 1999.

46. Mathews, J. and Fink, K., *Numerical Methods Using MATLAB*, Third Edition, Prentice Hall, 1999.

47. Fausett, L., *Applied Numerical Analysis Using MATLAB*, Prentice Hall, 1999.

Books on Fibonacci Numbers

48. Posamentier, A. S., *The Fabulous Fibonacci Numbers*, Prometheus Books, 2007.

49. Vajda, S., *Fibonacci and Lucas Numbers, and the Golden Section*, Dover Publications, 2007.

50. Vorobiev, N. N. and Martin, M., *Fibonacci Numbers*, Birkhauser, 2003.

51. Dunlap, R. A., *The Golden Ratio and Fibonacci Numbers*, World Scientific Publishing, 1997.

52. Olsen, S., *The Golden Section: Nature's Greatest Secret*, Walker and Company, 2006.

53. Livio, M., *The Golden Ratio: The Story of PHI, the World's Most Astonishing Number*, Broadway, 2003.

54. Koshy, T., *Fibonacci and Lucas Numbers with Applications*, Wiley-Interscience, 2001.

55. Dobson, E. D., *Understanding Fibonacci Numbers*, Traders Press, 1994.

56. Hoggatt, V. E. and Bicknell, M., *Primer on Fibonacci Numbers*, Fibonacci Association, 1972.

Web Links for Fibonacci Numbers

57. Fibonacci Numbers at Wolfram MathWorld
http://mathworld.wolfram.com/FibonacciNumber.html

58. Fibonacci Numbers at Wikipedia
http://en.wikipedia.org/wiki/Fibonacci_number

59. Fibonacci Numbers and the Golden Section
http://www.maths.surrey.ac.uk/hosted-sites/R.Knott/Fibonacci/fib.html

60. The Mathematical Magic of Fibonacci Numbers
http://www.maths.surrey.ac.uk/hosted-sites/R.Knott/Fibonacci/fibmaths.html

61. The Fibonacci Sequence – Math is Fun
http://www.mathsisfun.com/numbers/fibonacci-sequence.html

62. Generalizations of Fibonacci Numbers from Wikipedia
http://en.wikipedia.org/wiki/Tribonacci_number

63. Random Fibonacci Sequences from Wikipedia
http://en.wikipedia.org/wiki/Viswanath%27s_constant

64. The Golden Mean and the Physics of Aesthetics by Subhash Kak
http://arxiv.org/abs/physics/0411195

65. Stepping Beyond Fibonacci Numbers – from Ivars Peterson's MathTrek
http://www.maa.org/mathland/mathtrek_09_30_02.html

66. Viswanath's Constant from PlanetMath
http://planetmath.org/encyclopedia/ViswanathsConstant.html

67. Divakar Viswanath, "Random Fibonacci Sequences and the Number 1.13198824...", Mathematics of Computation, Vol. 69, 2000, pp. 1131-1155.

Installation of MATLAB

In this book, it is assumed that you have already installed MATLAB on your computer system. To go through the eleven chapters of the book, you need to have MATLAB running on your computer. For help on installing MATLAB on your computer, check the following web links:

MATLAB Installation for Windows:

```
http://www.itc.virginia.edu/research/matlab/
     install/installwin.html
```

MATLAB Installation for Linux:

```
http://www.freebsd.org/doc/zh_TW/books/handb
     ook/linuxemu-matlab.html
```

MATLAB Installation for Macintosh:

```
http://www.itc.virginia.edu/research/matlab/
     install/installmac.html
```

MATLAB Installation for Other Systems:

```
http://shum.huji.ac.il/cc/matlR13linuxhomein
     st.html
```

MATLAB Guide to Fibonacci Numbers and the Golden Ratio

A Simplified Approach, for Students and Scientists and Engineers

MATLAB Guide to Fibonacci Numbers and the Golden Ratio

A Simplified Approach, for Students and Scientists and Engineers